MENSTRUATION

Best wishes Bob!

Thad Doggett

MENSTRUATION

Origin and Evolutionary Significance

Thad H. Doggett, M.D., M.S.

VANTAGE PRESS
New York

FIRST EDITION

All rights reserved, including the right of
reproduction in whole or in part in any form

Copyright © 1995 by Thad H. Doggett, M.D., M.S.

Published by Vantage Press, Inc.
516 West 34th Street, New York, New York 10001

Manufactured in the United States of America
ISBN: 0-533-11240-0

Library of Congress Catalog Card No.: 94-90410

0 9 8 7 6 5 4 3 2 1

To my wife, Gertrude,
and my children,
Conrad, Stephen, and Elisa,
for their patience and tolerance
while this book was being written

Contents

List of Illustrations   xix
Foreword by Raphael S. Good, M.D.   xxi
Preface   xxiii
Acknowledgments   xxvii

Introduction   1

Part One: Biologic Roots of the Human Reproductive Cycle

I. Origin of the Estrous Cycle   9

    1. Evolution of menstrual cycle from estrous cycle. **9**
    2. Estrus defined and its purpose. **9, 27**
    3 (a). Behavior in estrous cycle same as behavior in menstrual cycle, only bleeding of menstruation differentiates the two cycles. **10** (b) Evolutional path of behavioral change differs from path that produces menstruation although both are physiologically produced by same endocrines. **10**
    4. Estrous or menstrual cycle in the human is presently undergoing evolutional change. **10**
    5. Evolutional change in estrous cycle is evolutional change in placentation. **11**
    6. Evolutional change in placentation that previously advanced certain lower primate lineages to their present higher primate status produced menstruation in upper primates. **11**
    7. Development reached by brain and central nervous system determines rank of a form in mammalian hierarchy. **11**
    8. Advancing placentation physiology advances central nervous system development by advancing embryonic respiration. **11**
    9 (a). Estrous cycle of mammals and birds arose from reptilian reproductive cycle. **12** (b) Birds continued the reptilian reproductive pattern of hatching large-yolked eggs. **12** (c) Mammalian lineage arose through marked evolutional change in the reproductive pattern that established the estrous cycle. **13**
    10 (a). First change toward establishing the mammalian estrous cycle was

true anatomical reduction in ovarian egg size. **14** (b) Mammalian egg continued to become smaller, finally becoming a particle in placental mammals. **14**
11 (a). After maximum reduction in anatomical size, two new organs, the uterus and placenta, formed in reproductive cycle of mammals. **14** (b) Oviparity became converted to viviparity. **14**
12. Enlarging brain and central nervous system resulting from improved respiratory nutrition in placentation produced some change in mammalian body form. **14**
13. Development of the uterus and viviparity was accompanied by formation of the corpus luteum in the ovary that controlled uterine function. **15**
14 (a). Variation in primates shows many small confusing steps in the evolutionary path of estrous cycle to menstrual cycle. **15** (b) Comparison of reproduction in mammalian forms over a longer evolutionary time span can reveal the evolutionary path more clearly. **16**
15. Establishing steps in an evolutionary path requires comprehension of the natural forces that selected them. **16**
16 (a). Estrus in reptiles resulted in coitus and storage of sperm in female body, while the reptilian reproductive cycle involved the laying of fertilized eggs at a later time. **16** (b) Higher body temperature developing in mammals precluded stored sperm. **17** (c) Merging of reptilian coitus with egg production (ovulation) established the estrous cycle in mammals. **17**
17. Estrus occurs earlier in mammalian reproductive cycle when a tiny egg is produced. **17**
18. Reproduction in reptiles advances to mammalian reproduction on same evolutionary path reptiles advanced from amphibians, greater oxygen consumption in metabolism, and increase in body heat. **18**
19. Large-yolked reptilian egg is part of reptile's adaptation to dry terrestrial environment. **18**
20. By continuing the reptilian reproductive pattern, birds have not been able to advance development of the reptilian brain and central nervous system that has been accomplished through mammalian placentation. **19**
21. Reptile hatchling was more a miniature adult, while larger brained mammalian hatchling required nurturing. **19**
22. Mammalian lineage began incubating the egg internally. **19**
23. Reptile abandoned clutch after laying eggs ending the reproductive cycle, while the cycle ends in mammals and birds after hatchling is nurtured. **18**
24. Loss of large egg reinstitutes reproductive cycle in birds, while loss of tiny mammalian egg also reinstitutes reproductive cycle but loss is not detected through organs of special sense. **20**
25 (a). Tiny mammalian egg becomes endocrine organ. **21, 22, 25, 26, 27** (b) Its loss is detected in loss of its endocrine support of all other endocrine control of the uterus. **22** (c) Without endocrine support from the egg, the uterus

of the upper primate empties, ending the reproductive cycle. **22** (d) Uterine-emptying or delivery functions in the mammalian uterus replaced egg-hatching functions in hard-shelled egg. **23** (e) Soft membranes of uterine egg at term in placental mammals are hatched or opened during its delivery. **23**

26. No offshoots from transitional forms between reptiles and mammals that could show past evolutionary steps have survived. **22**

27 (a). Internal incubation of mammalian egg brought it into environment of unlimited nutrition. **22** (b) Mechanical support from uterus supports development and growth of the mammalian egg to comparatively huge size before hatching and delivery. **23**

28. Internal morphology of huge mammalian egg at term has important role in uterine delivery mechanics. **23**

29 (a). Eggs of aplacental mammals do not affect maternal physiology. **23** (b) Ovary carries the reproductive organs through reproductive cycle, incubating, and delivering eggs, even though they may be unfertilized and dead. **23, 24**

30. Lack of nursing embryos in pouch of aplacental mammals institutes a new reproductive cycle. **24**

31. Lack of nursing neonate also institutes new reproductive cycle in placental mammals. **24**

32. Most frequent cause of absent neonate in reproductive cycle of mammals is failure of egg to become fertilized. **25**

33. Shift of embryo control of the reproductive cycle from after delivery to before delivery has occurred in placental mammals with lengthier intrauterine gestation periods. **26**

34 (a). Death or loss of egg or embryo initiates uterine-emptying physiology in placental mammals. **25** (b) Uterus empties when egg dies irrespective of whether it has developed to term or has failed to become fertilized. **26**

35 (a). When reproductive cycle ends from failure of egg fertilization to occur, the cycle is designated an estrous cycle. **27** (b) When cycle ends after fertilization has occurred, it is designated either abortion or premature or term delivery. **27** (c) In all instances uterine-emptying physiology is the same. **27** (d) Redefinition of the estrous cycle should be considered. **27**

II. The Estrous Cycle and Reproduction in Infraprimate Mammals   30

1. Gradual lengthening of estrous cycle in aplacental mammals began with evolutionary changes within the egg. **30**

2 (a). Eggs of aqueous vertebrates are small with little yolk. **30** (b) Loss of yolk in internally incubated egg of mammalian lineage returned it to aqueous environment. **31** (c) Aqueous environment of maternal tissues provided unlimited nutrition. **31** (d) More oxygen available from maternal tissues fos-

tered larger brain development in placental mammals. **31**

3 (a). Large-brain development requires large embryo or fetus. **34** (b) Central nervous system controls regulation of body temperature. **32**

4 (a). Central nervous system regulates body temperature in controlling heat loss of higher body temperature. **32** (b) Body morphology is also a factor in heat loss. **32**

5. Larger brain formation during embryogenesis requires that oxygen supply keep up with brain growth and enlargement. **33**

6 (a). Yolk is poor source of both oxygen and respiratory surface. **34** (b) Birds with large-yolked eggs advanced reptile brain very little as large yolk mass in egg was barrier to early oxygen supply in embryogenesis. **33** (c) Allantoic respiratory surface in bird egg begins to absorb more oxygen late in embryogenesis. **33** (d) Musculoskeletal system, which develops after central nervous system is established, makes greater use of late availability of oxygen in bird embryogenesis. **33**

7 (a). Loss of yolk and increasing egg membrane surface in placental mammals advanced central nervous system development seen in existing mammalian orders. **34** (b) Mammalian egg in placental mammals establishes greater respiratory surface with endometrium. **35** (c) Greater egg surface and embryo morphology altered emptying function of uterus. **38** (d) Aplacental mammals open or hatch egg during delivery. **36** (e) Uterine-hatching and delivery functions more orderly in placental mammals. **37**

8 (a). Aplacentalia were dominant mammalian groups in the past. **37** (b) Brain and central nervous system of Aplacentalia smaller than more recent placental mammals of modern times. **37** (c) Hind parts of Aplacentalia, like reptiles, tend to grow larger and more developed than foreparts. **38**

9 (a). Lowest of existing placental mammals is the pig. **38** (b) Morphology and development of foreparts resemble Aplacentalia. **38** (c) Reproduction of pig more advanced but resembles reproduction of Aplacentalia. **38** (d) Eggs of marsupials lightly adhere to endometrium, while larger chorion of pig also only lies against endometrium. **38**

10 (a). Uterus of Aplacentalia is a pair of bilaterally separate organs that function completely independent of each other. **39** (b) Uterus of pig is a single-chambered organ of two horns. **39** (c) In acquiring their spacing in uterine horns, pig eggs "migrate" from one horn to the other. **39** (d) Embryogenesis is markedly delayed in the pig as the chorion spreads its surface over the endometrium. **39** (e) Vascularization of the chorion shifts from the diminishing yolk sac mesodermal surface to the mesodermal surface of the allantoic pouch enlarging from metabolic waste storage. **40**

11 (a). Nutrition passes from endometrium into pig eggs. **40** (b) Questionable whether substances pass backward from pig egg into maternal circulation as pig egg stores metabolic waste. **41** (c) Degeneration of the several

corpora lutea in the pig ovary probably brings on uterine emptying at term as well as at end of estrous cycle. **41** (d) Reproductive cycle of pig, as in marsupials, may well be entirely under control of the ovary. **42**

12. Delivery of spaced pig fetuses more orderly than the haphazard delivery of the "bunched" eggs containing embryos in marsupials. **42**

13. Involution of sow uterus involves absorption of redundant and collapsed endometrium without any bleeding or lochia. **43**

14 (a). Most existing ungulates are forms from slightly higher advanced offshoots than the pig. **43** (b) Head and neck developed more in higher ungulates, and neonate with longer legs can stand almost immediately at birth and run with the herd. **43** (c) Litter reduced to single egg in which fetus develops in one uterine horn while membranes spread to both horns for greater surface. **44** (d) Single corpus luteum could not sustain such a large uterus during gestation. **44** (e) Ovulation and corpus luteum formation occur throughout pregnancy in some (mare), in others (cow) estrogens and progestins produced by "placentomes" in egg membranes. **44, 45**

15. Mechanical problem during parturition of higher ungulates can occur from the poor utero-fetal morphology. **45**

16 (a). Cervix and cervical function are present in the delivery mechanism of single fetus ungulates. **45** (b) A long umbilical cord permits delivery of the fetus while membranes continue to function. **45** (c) Delivery of the huge and extensive placenta can be complicated and last over several days. **45** (d) Involution of the uterus separates the placentomes of the egg membranes. **45** (e) Lochia may become bloody from involution of endometrial caruncles. **46**

17 (a). Placenta of Carnivora attaches earlier and deeper in a band or zone, establishing more effective and earlier increase in respiratory nutrition for the egg. **47** (b) Egg does not penetrate endometrial blood vessels, consequently a fetus in the litter higher attached in the uterine horn can be delivered over placental site of a previously delivered fetus that attached lower in the uterine horn. **48, 49** (c) There is no cervix in Carnivora uterus as litters rather than single fetuses are delivered. **49**

18 (a). Separation and delivery of the placenta in all infraprimate mammals does not involve bleeding. **50** (b) Uterine functions end estruos cycle by absorption of damaged endometrium through leukocytosis without bleeding. **50**

III. Development of Menstruation within the Estrous Cycle of Primates   52

1. Evolutionary path of reproductive functions in primates is not a straight path but a very irregular "zig-zag" course. **52**
2. Evolutional advancement of reproductive physiology in primates centers

about advancing hemotrophic nutrition in early stages of embryogenesis. **52**
3. Evolutionary path of primate reproduction studied easier in larger primate groups more distantly related. **53**
4 (a). Mesodermal structures develop precociously in the primate egg. **53** (b) Maximum use of body oxygen source has been established in implantation, embryogenesis, and fetal growth of the human egg. **53** (c) Further evolutional advance of human reproduction requires changes in adult structure and physiology. **54**
5 (a). Comparatively huge central nervous system in the fetus makes oxygen delivery most limited nutritional factor in human reproduction. **54** (b) Oxygen is not stored in nervous tissues. **54, 55** (c) Constant uninterrupted oxygen supply essential in human reproduction. **55**
6 (a). Fusion of Muellerian tubes into uterus formation first began in ungulates to increase egg-membrane surface. **55** (b) As egg began to attach deeper to endometrium, greater oxygen source was reached. **55** (c) Further egg membrane surface became unnecessary and further fusion of Muellerian ducts ceased. **55**
7 (a). Environment of internally incubated egg is genetically determined. **56** (b) Fusion of Muellerian ducts is also genetically determined. **56** (c) Adaptation of internally incubated mammalian egg during early embryogenesis is adaptation to its own later developed body environment that is determined by later expression of its own genotype. **56**
8 (a). Uterus is not primarily a nourishing organ. **56** (b) Human egg possesses all nourishing structure and function in placenta formation as exemplified in ecotopic pregnancy. **56** (c) Primary function of mammalian uterus is to deliver the egg—a mechanical function and accomplishment. **57**
9. Eggs of pig sit atop endometrium with occluded but actual cavity space existing between surfaces of chorion and endometrium. **57** (b) Chorion of single fetus ungulates erodes endometrial epithelium to form caruncles and lengthen incubatory period of gestation. **57** (c) Carnivora eggs invade endometrium sooner and deeper, leaving chorion of the egg poles that are in contact with uterine lumen bare of villi. **58**
10 (a). Eggs of lower primate reduced to a single ovum as in higher ungulates. **57, 61** (b) Greater oxygen demand of egg brought on further fusion of Muellerian ducts to complete formation of a single-chambered uterus. **58, 61, 62, 63** (c) Egg invades endometrium progressively deeper, entering maternal blood vessels establishing the ultimate respiratory surface. **59** (d) Anterior and posterior surfaces of the endometrium thicken into cushions designated "placentoids" for egg nutrition. **58** (e) The two endometrial surfaces produce a bilobed placenta. **58** (f) Like poles of Carnivora egg areas of smooth avillous chorion contact uterine lumen. **58, 65** (g) Trophoblast of egg in lower primates goes through shorter histotrophic stage than egg of

infraprimate mammals. **59** (h) Cords of trophoblastic syncytium with mesodermal cores penetrate endometrial blood vessels. **60** (i) Tiny lacunae grow into labyrinthine channels forming a trabeculate placenta. **60** (j) No bloody placental sign or anovulatory bleeding occurs in this group of primates. **60** (k) Menstruation, when it occurs, is microscopic, revealed as a layer of red cells in centrifuged vaginal washings. **58** (l) Uterine squeeze of parturition produces only stromal damage of the endometrium. **59**

11 (a). Ovum of upper primate establishes hemotrophic nutrition from the beginning as it embeds. **60** (b) Egg invades deepest while still a blastocyst. **64** (c) Egg invades only one endometrial surface, establishing a single lobed placenta. **64** (d) Egg completely penetrates endometrium with formation of a capsularis and a discoid placenta. **64** (e) Endometrium does not form placentoid cushions; all endometrial surface is same thickness. **60** (f) Functional (deciduous) layer of endometrial surface develops over the basal layer of endometrium attached to muscle wall of uterus. **60** (g) Uterus delivers superficial layer of endometrium containing the fixed egg rather than the egg itself. **67** (h) Uterus separates and delivers the superficial endometrial layer irrespective of egg fertilization. **68** (i) If egg is not fertilized and does not attach to the endometrium, the cycle ends in menstruation and the cycle is designated a menstrual cycle. **69** (j) If fertilized egg implants, cycle ends as an abortion, premature delivery, term birth, or stillbirth. **68**

12 (a). Uterine squeeze that separates the two endometrial layers staunches the arterial blood supply to the superficial separating layer. **68** (b) Control of hemorrhage at the plane of cleavage is through a special arteriolar blood supply to endometrial surface capillaries. **68** (c) Muscular sphincters at the base of the arterioles clamp off the copious arterial blood supply to the endometrial surface. **68** (d) Muscular sphincters controlling blood supply to endometrial surface belong to inner muscle layer of uterus, making staunching of the endometrial surface blood supply simultaneous with its separation and delivery. **68** (e) Both staunching and delivery are accomplished by the same uterine squeeze. **68** (f) The uterus carries out this same performance during menstruation early in the reproductive cycle, at term during parturition, or any other stage of the reproductive cycle following ovulation or production of the egg. **69**

Part Two: Physiologic Mechanics of the Human Uterus

IV. Function of the Human Uterus before Labor 73

1. Primary functions of the ovary and oviduct in reproduction are clearly understood, but functions of uterus are not so clear and include enigmas such as menstruation and dysmenorrhea. **73, 74**

2. Although splendid progress has been made in reproductive endocrinology, childbirth remains a mechanical accomplishment not fully understood. **74**

3 (a). Mechanics of childbirth include morphology and geometry of egg contents at term as well as morphology and geometry of the uterus. **74, 75** (b) Blood circulation and oxygen delivery to the placenta must be maintained during application of stressful mechanical pressure. **74**

4 (a). Incompressible amniotic fluid distributes and diverts mechanical pressure of the uterus. **75** (b) Uterine pressure mechanically supports and keeps intact egg membranes and egg contents as term approaches. **79** (c) Human embryo develops within three separate enclosures of fluid, amniotic enclosure that is within the chorionic enclosure located within the fluid-filled uterine lumen. **78** (d) The fluid-filled compartments become a single fluid compartment responding to uterine pressure as the enclosing membranes fuse. **79** (e) The single fluid compartment is developmentally located within the superficial deciduous layer of endometrium although it becomes mechanically located within the uterine cavity. **77, 78** (f) Uterine cavity is an actual cavity containing a flow of glandular secretion at beginning of gestation but becomes obliterated and is reduced to a potential cavity during last half of gestation. **78**

5. Amniotic fluid circulates, chorionic fluid is displaced, and glandular secretions of uterine cavity flow and discharge in early stages of human gestation. **78**

6 (a). Space for growing fetus is enlarged by increasing amounts of amniotic fluid that distends uterus without mechanical effect on fetus and cord suspended within. **80** (b) Accumulation of distending amniotic fluid has marked mechanical effect on placenta formation and function. **80**

V. Function of the Human Uterus during Labor 84

1 (a). Uterus with its contents, the egg and endometrium, all grow and enlarge volume and surface during gestation. **84** (b) Growth and enlargement is reversed during involution, reducing uterus to premenstrual size. **85**

2. At any particular moment uterine muscle cell or fiber exists in only one of three states, growing and enlarging, shrinking or involuting, or resting between the other two states. **85**

3 (a). Two types of uterine contractions, intermittent uterine contractions that persist throughout gestation, and shrinkage or involution of uterine musculature. **85** (b) Both types of contraction occur simultaneously during labor. **85** (c) During estrogen stimulation and growth each muscle cell is longer after intermittent contraction. **87** (d) During labor or involution each muscle fiber is shorter after an intermittent contraction. **87**

4 (a). Involution of uterine musculature during labor and delivery (parturition) is not generalized like intermittent contractions. **87** (b) Before labor there is equal squeeze over entire uterus. **87** (c) Contraction and compression move parallel to venous channels toward uterine hilus; venous channels are not sphincterized, and contraction of muscle fibers tends to pull venous channels open. **88** (d) Labor begins with involution of musculature associated with the cervix and cervical effacement. **89** (e) Cervical dilation requires pressure of fetal presenting part. **97** (f) Cervix is a "knob" of uterine tissue that converts to uterine cavity surface during effacement. **87**

5 (a). Effacement of cervix begins at internal os as a point on the decidua capsularis that seals the opening. **89** (b) As cervix effaces, the point is pulled into expanding ring of fused decidua vera and decidua capsularis. **89** (c) The stretch of the underlying decidua capsularis containing its attached chorion ruptures or bulges through cervical os. **97**

6 (a). Rupture or ballooning of egg membranes through effacing cervix and dilating internal os adds surface to the uterine cavity while reducing its volume. **97** (b) Uterine squeeze becomes applied to fetal presenting part. **97** (c) Fetal presenting part acts as a fulcrum or wedge similar to a stopper inside a bottle of fluid. **97** (d) Uterine pressure shifts from a generalized pressure over entire egg surface to anterior surface of egg membranes covering fetal presenting part. **97** (e) Shift of pressure is due to retained hindwater as circular or binding squeeze of uterus shifts posterior to fetus. **97**

7 (a). Squeeze and pressure of hindwater keeps placenta distended and functioning during labor. **98** (b) Dilation of effaced cervix dilates and shortens upper vagina, bringing fetal presenting part to introitus without shift of fetal position within the uterus. **99** (c) Uterus containing fetus moves downward into pelvis. **99** (d) Uterine musculature shifts around fetus without changing fetal location within the uterus. **99**

VI. Function of the Human Uterus during Birth of the Fetus   100

1 (a). Delivery of the fetus is expulsion of the fetus live or dead while birth of a fetus is shift of respiration from placenta to fetal lungs. **100** (b) Opening egg membranes may require hours, while hatching or delivery of fetus is instantaneous. **101** (c) Live fetus before labor contributes to circulation of amniotic fluid. **101, 102** (d) Oxygen and respiratory gases are chemically bound; no free gas exists in pregnant uterus with living embryo or fetus. **102** (e) Desiccation of fetus and membranes begins when they reach atmosphere. **101, 102** (f) Endometrial covering over maternal surface of delivered egg is not grossly visible. **102** (g) Egg grows and develops in its own folded and packaged membranes that are opened and "unpacked" during parturition. **102**

2 (a). Human egg at term is essentially a single bipolar membrane. **104** (b) Fetal pole enlarges in three dimensions while placental pole expands in two dimensions. **104** (c) Birth is switch of respiration, nutrition, water absorption, and excretion from attached two-dimensional surface pole to the freely suspended and motile fetal mass. **105** (d) Switch in nutrition from one end of egg membrane to the other follows in accordance with the mechanical stages of parturition that are determined by morphology of egg contents. **106** (e) Squeeze of uterine involution is divided into stages by morphology of egg contents when egg becomes compartmentalized after its membranes are opened. **108**
3 (a). Fetal morphology is such that when uterine squeeze begins to close off placental circulation, fetal nares have reached the atmosphere. **106, 107** (b) Axis of fetus lengthens as uterine musculature extrudes the fetus. **106** (c) Single irreversible intermittent contraction of uterus delivers fetus simultaneous with shut down of maternal circulation to the placenta. **108**
4 (a). Uterine squeeze that expels fetus also expels anterior cavity surface. **109** (b) Collapsing and contracting uterine cavity surface is extruded into vagina as fetus exits, reestablishing cervical volume and obliterating cavity volume. **110**

VII. Function of the Human Uterus after Delivery of the Fetus   111

1. Uterine function following delivery of fetus is the most critical stage of parturition and is climax of evolutionary path of human reproductive physiology. **111**
2 (a). Live birth at every parturition is not necessary to survival of human species. **111** (b) Although live fetus not essential, delivery of every placenta is, and it must never fail. **112**
3 (a). Placenta and membranes constitute remains of egg that must be delivered. **113** (b) Egg membranes are inseparably attached to deciduous endometrial layer; consequently endometrial layer containing egg remains is separated and delivered by uterine squeeze. **115** (c) Final uterine squeeze of parturition gathers endometrial surface layer and its contents into a volumetric mass for separation and delivery. **126** (d) Mass consists of contracted placenta and egg membranes within endometrial surface layer. **126** (e) Final uterine squeeze contracts subplacental muscular layer. **126**
4 (a). Contracted placental membrane locks into volume with fixed surface greater than its attachment. **116** (b) As placental membrane expands during formation with increase of its distended surface, secondary villi add tissue volume, rendering the membrane irreversibly expanded. **123**
5. Cleavage occurs in the capillaries linking the deciduous superficial layer with the capillaries of the basal layer. **119**

6. As endometrial surface containing placenta is separated by subplacental muscular contraction, the copious arterial blood supply to the severed surface capillaries from spiral arterioles is staunched. **124, 125** (b) The contracted endometrial surface is rendered ischemic as it is pulled into its position lining the occluded uterine cavity space. **117**

VIII. Function of the Human Uterus between Pregnancies   **127**

1 (a). Contraction and involution of uterus continues two to three weeks postpartum. **127** (b) Postpartum cramps can be very painful. **127** (c) While nursing, the mother may feel weightiness in pelvis from contracting organ. **127** (d) After approximately ten days, uterus approaches its prepregnancy size in resuming sperm transmitting and menstrual functions. **127**
2 (a). Uterus immediately postpartum weighs over two pounds and diminishes to sixty grams or so in ten to fourteen days. **127, 135** (b) Involuting uterus returns nitrogen, water, and minerals to maternal circulation. **127** (c) Uterine cavity remains obliterated and its surface held ischemic for at least an hour. **127, 128** (d) Autopsied specimens at this time reveal little evidence of placental location on the endometrial surface as capillary surface of the endometrium containing the placenta has been removed. **128**
3 (a). Endometrial arterioles both coiled and straight do not enlarge during gestation. **128, 129** (b) Coiled arterioles lengthen to continue blood supply to placental capillaries (sinuses) as endometrial surface spreads during placenta expansion. **129** (c) Opened ends of coiled arterioles bleed intermittently as capillary growth and resurfacing begins an hour or more after placenta is delivered. **130** (d) Healing and resurfacing of the endometrial cavity would be complete in probably twenty-four hours or so were it not for continued uterine involution and sloughing of healed but diminishing endometrial surface. **133**
4 (a). Bleeding and sloughing of new-formed endometrial surface forms lochia and is emptied as discharge. **133** (b) Uterine lumen becomes reestablished in accommodating discharging lochia until involution of musculature is complete. **134, 135**

Part Three. Mechanism of Menstruation

IX. Delivery Performance of the Human Uterus before Term   **139**

1. Normal or ideal delivery of the uterus is delivery of an egg with fully developed morphology. **139**
2 (a). Altered body morphology in postterm fetus interferes with pattern of normal uterine mechanisms during parturition. **140** (b) Altered morphol-

ogy of fetus also impedes mechanics in premature parturition. **141, 145**

3 (a). Uterine mechanics are influenced by ratio of fetal size to placental surface. **142, 144** (b) Ratio is smaller in cases of premature parturition. **142, 144** (c) Incarceration of smaller fetal mass by circular binding musculature more likely in premature parturition. **146**

4 (a). In cases of undisturbed and unmolested premature parturition, uterine-emptying mechanics attempt to follow same pattern they go through in ideal term parturition. **144, 145, 146** (b) Subplacental muscular layer carrying placental blood supply covers greater portion of cavity area. **144** (c) Fluid in intact amnion protects embryo and fetus from traumatic uterine pressure. **146** (d) Entire unhatched egg may be delivered intact, i.e., egg opening or hatching mechanics may not operate because of fetal/placenta ratio and egg-membrane elasticity. **141** (e) Unhatched premature egg like full-term egg is delivered embedded within its endometrial surface. **141**

5 (a). In gestations of only four to six weeks, parturition functions of the uterus empty only the redundant endometrial membrane containing the blighted ovum. **147** (b) Embryo or fetus is absent and egg exerts no mechanical influence whatsoever on delivery mechanics. **148**

6 (a). Still earlier premature involutions of the uterus produce what appears to be delayed, heavier than usual, irregular menstrual periods with clots. **147, 149** (b) Previous pregnancy test may confirm diagnosis of microabortion. **147, 148, 149** (c) Involution and emptying of uterus after egg has implanted in endometrium is conventionally defined as an abortion. **149** (d) Involution and emptying of the uterus before egg implants or in absence of egg implantation is conventionally defined as a menstrual period. **153** (e) Involution and emptying of the uterus as the egg implants cannot be defined; uterine-emptying physiology is identical in both menstrual cycle and parturition. **153** (f) Fertilization, the actual entry of sperm into the egg, produces no effects on maternal reproductive functions. **152** (g) Results of successful fertilization are continuation and furthering the existing reproductive physiology. **152, 153** (h) Uterine-growth physiology initiated in beginning of the human reproductive cycle is balanced by involution of the uterine-emptying physiology between menstrual periods. **154**

Bibliography

List of Illustrations

4.1. Representation of the solid, or formed, egg contents of the human uterus delivered at term. **76**
4.2. The human egg at term within the uterus before labor begins. **77**
4.3. Diagram of the human uterus in the third month of pregnancy demonstrating the formation and relationship of fetal and maternal membranes. **78**
4.4. Main structures of the human endometrium prior to egg implantation as seen in tissue sections prepared for microscopic study. **82**
4.5. Diagram of a section through the intact human placenta near its junction with the umbilical cord. **83**
5.1. Diagram indicating the two functional states of uterine musculature—growth and involution. **86**
5.2. Pregnant uterus at term isolated from its attachments, showing direction and effect of the mechanical force exerted by an intermittent uterine contraction before onset of involution (labor). **88**
5.3. The three ligaments anchoring the uterus to the bony pelvis during full cervical dilation. **90**
5.4. Diagram of fully dilated cervix. **92**
5.5. Diagram of effacing cervix dilating and adding to inner surface area of the uterine cavity. **93**
5.6. The placenta maintained in its distended and functional state by the pressure of the hindwaters in the compartmentalized egg. **94**
5.7. Direction of uterine pressure when the cervix is partially (a) and fully (b) dilated. **95**
5.8. Diagram representing a sagittal section through the cervix and uterus before labor (a) and as the cervix effaces (b). **96**
6.1. Scheme of the developing egg membranes in mammals. **103**
6.2. The contracting and involuting muscle bundles thickening the uterine wall in its extraplacental regions. **107**
7.1. Diagram of a segment of uterine wall with corresponding stretched and functioning placental segment attached (a) and contracted (b). **116**
7.2. Diagrams of extruding cord after fetal delivery. **117**
7.3. Mechanical forces exerted by the muscular wall of the uterus in separation and delivery of the placenta. **120**

7.4. Diagram showing flattening of radial structure. **122**
7.5. Diagram of advance of secondary villi. **123**
7.6. Diagram of a segment of uterine wall and endometrium beneath the placenta (a) and contracted after delivery of the placenta (b). **125**
8.1. Diagram showing contracted uterine wall (a) and sloughing of endometrial surface (b). **131**
8.2. Different stages in the healing and repair of the surface of the endometrial cavity. **134**
9.1. The intact decidua compacta containing the intact premature egg. **141**
9.2. Diagram of delivered extrafetal portion of the membranes filled with water. **143**
9.3. Diagram of a hemisected preterm uterus with the removed portion of the uterus dissected off the intact placenta. **145**
9.4. Delivery of an early intrauterine pregnancy of only a few weeks. **148**
9.5. Two stages of the delivery functions of the uterus presented in figure 9.4 but in the absence of an implanted egg (menstruation). **151**

Foreword

Dr. Doggett's monograph represents the culmination of his many years of interest in understanding the development of the menstrual cycle in humans. He has formulated a masterful synthesis of data, some of which is well known to various disciplines and mostly unknown to other disciplines to whom this work is directed. As an example, the physiology of the hatching of hard-shell eggs may be common knowledge to ornithologists but completely unknown and thought of as of no significance to the practicing obstetrician. Similarly, the concerns of the obstetrician in the immediate delivery of the placenta following term birth of a human fetus may well be considered of no relevance to the ornithologist, anatomist, embryologist, or one interested in evolution. This monograph clearly delineates the relationship between such seemingly disparate concerns and describes the bridge between them. In addition to dealing with the embryology and physiology of seemingly unrelated phenomena in various species, the author beautifully demonstrates the adaptive value of these evolutionary changes. The anatomist, embryologist, Darwinian, or obstetrician-gynecologist who patiently studies this monograph will gain a remarkable understanding of the development and meaning of the menstrual cycle and will be able to answer the question, "Why do women menstruate?"

—Raphael S. Good, M.D.
Clinical Professor Psychiatry & Ob-Gyn
Univ. of Miami School of Medicine

Preface

In 1961, in an article in the *Journal of the Florida Medical Association*, I presented the mechanism of menstruation and its clinical significance. At the time clinicians did not grasp the significance of the concept that was being presented. Some textbook authors have since concluded that "menstruation is the parturition of failed fertility" (Cunningham et al., 1989, p. 35). Others have continued to state that the mechanism of menstruation is not clearly understood because functions of the endometrium have not been fully elucidated (Scott et al., 1990, p. 741). This is frustrating for clinicians having to manage disorders of unexplained uterine phenomena. Their reports on dysmenorrhea, menstruation, and other phenomena of uterine reproductive performance continued to refer to the unknown cause or biologic significance of each.

Some biologists, on the other hand, consider the concepts presented in that monograph as nothing scientifically new or novel but representing only a review of available data. Although scientific data necessary to understand menstruation have been presumably complete for half a century or more, any attempts to synthesize this information into a single concept that harmonizes all known facts have not been recognized by either clinicians or biologists. Such a synthesis necessarily requires some review of accepted knowledge, but interpreting and correlating data also requires discussion. This can be as important to a scientific investigation as reporting the data and may not be entirely philosophical, as is indicated in this case.

The lengthier and more detailed discussion of menstruation, including the evolutionary origins of its associated physiology presented in this book, is an effort to focus attention on specific areas that have been overlooked by readers of my article. Terminology, in some instances, may appear overly simplistic and even crude to a sophisticated investigator in molecular biology. An effort has been made to interpret some evolutionary principles for the clinician reader, and certain clinical terms have been converted to language considered by the author to be more easily interpretable by biologists. Hopefully, this will cause some attention from both groups to become focused on the problem.

Reproductive physiology is presented from the evolutionary standpoint in which environmental selection accounts for tissue and organ functions becoming altered and advanced in steps of adaptation. Morphology and evolutionary effects of morphological change, while currently *passé* among many investigators, is the method of study used here, the scientific explanation of menstruation, at the present time, requiring the utilization of this older approach. The material is unavoidably dense and fatiguing to the reader in some areas because of directional variations in the lengthy evolutionary paths taken by mammals and primates.

Critical functions of all uterine tissues, including the endometrium, are carefully detailed in terms of their relationships with each other and with the morphology and physiology of the developing ovum. All functions of the oviduct and uterus are under control of the ovum. As the human ovum begins to mature in the ovary, it establishes its own endometrial bed in the uterus. This endometrial bed is the decidual endometrial surface that provides the ovum with a continued source of nourishment during its intrauterine development. Still more important for humans, however, this membrane additionally provides a mechanical means of safely separating the dangerously intimate human egg from its mother's nourishing blood circulation. In pregnancy, the human egg fixes inseparably within this membrane, which is later separated and delivered during labor by the action of the uterus. The uterus also performs this same function (separating and delivering the endometrial surface) in the instance of the menstrual cycle, in which case the infertile ovum has failed to attach itself within the membrane surface. The uterus thus performs the same delivery function in the menstrual cycle that it performs in labor at term or at any other stage of the reproductive cycle. This understanding is critical to the clinician seeking clarification of enigmas in obstetrics and gynecology, which include menstruation, dysmenorrhea, PMS (premenstrual syndrome), and the premenstrual stage of endometrial ischemia. Lack of such an understanding accounts, to some extent, for the arbitrary subdivision of this field of medicine into obstetrics on the one hand and gynecology on the other.

In the human, the endometrium has evolved into two membranes with different functions. Stages in their evolution from a single adeciduous membrane can be seen in living mammals and are discussed in reference to their latest stage in the human. Separation of the membrane surface in humans necessarily includes staunching its intense blood supply, whether it is in its minimal stage in the menstrual cycle before menstruation or at its maximum stage during parturition. The staunching is the result of a muscular squeeze of the uterus that renders ischemic the deciduous endometrial surface where the egg fixes and the placenta attaches. Circulation to the basal layer adjoining the muscularis is left intact. The intense uterine squeeze also accounts for dysmenorrheal pain, embryo and fetal delivery, as well as control of endometrial hemorrhage in the third stage of labor. Delivery and staunching functions are accomplished by the same uterine contraction because of the special morphological arrangement of uterine musculature about the arteriolar bases of the separate arteriolar system supplying only the endometrial surface.

This complex of emptying function and structure in the uterus of higher primates gradually evolved. It began with oviductal function of oviparity in premammalian ancestry and further evolved through ovoviviparity and viviparity to reach the stage of euviparity in upper primates and humans. Euviparity, in which embryonic respiration is further enhanced by discharge of all embryonic waste into the maternal circulation, accounts for the enormous intimacy of the human egg and maternal circulation.

This evolutionary path in mammals is a continuation of the evolutionary path in terrestrial vertebrates that has led to progressively larger brains. The richer arterial circulation supplied to the ovum in the mammalian reproductive cycle provides more oxygen for evolutional enlargement of the embryonic and fetal brain during embryogenesis, but staunching and ischemic control of the increasing endometrial vascularity, preventing hemorrhage during egg hatching and delivery, necessarily became an accompanying evolvement. This evolutionary development in placentation and uterine function accounts for the enlargement of the brain in primates and the sudden appearance of the human family that is revealed in the fossil record.

These biologic roots of the menstrual cycle are presented in the first three chapters, followed by a short chapter summary that may prove helpful to the reader. In the next five chapters, details of each mechanical stage of uterine emptying in the human are covered in its own separate chapter. The final chapter presents the emptying function of the upper primate uterus at progressively earlier stages of ovum development until the mechanism of menstruation is reached and revealed in the earliest stage.

Acknowledgments

I wish to express my appreciation to Dr. William Little, chairman of the Department of Obstetrics and Gynecology, University of Miami, and Dr. Frederick S. Hulse, professor of anthropology, University of Arizona, for their suggestions and criticisms in early revisions of the manuscript. Dr. Raphael Good of the Departments of Psychiatry and Obstetrics and Gynecology, University of Miami School of Medicine, provided invaluable suggestions for the present revision.

MENSTRUATION

Introduction

Research of the past fifty years has accounted for rapid progress in reproductive endocrinology, resulting in a broader range of therapeutic modalities for clinicians managing aberrations of menstruation. Very little progress, however, has been made in understanding menstruation. The specialty practice of obstetrics and gynecology has always been deeply involved in the management of menstruation and its "disorders," but just how profound the involvement is cannot be known as long as menstruation is not understood sufficiently to establish its true definition and scope. At our present level of understanding, for example, the statistical average, but not the medical normal or ideal amount of menstrual hemorrhage, can be determined. Every woman, from her menarche on, is left to monitor her monthly blood loss and judge what her own normal amount of bleeding should be.

Data on menstruation have accumulated in the past as a result of rapid progress in several disciplines. Anatomy, physiology, pathology, embryology, biochemistry, and pharmacology provided important contributions to the present state of our knowledge. At the present time, rapidly expanding molecular biology research is adding an even greater amount of data. Research efforts now focus on acquiring information at the molecular level, with less attention given to the broader interpretation of the data as it relates to the explanation of menstruation. The correct explanation of menstruation, consequently, continues to remain elusive. Clinicians, in the meantime, are left committed to managing aberrations of this phenomenon without comprehension of its true biologic significance.

An up-to-date reference work on the problem of explaining menstruation is not forthcoming, since there is presently little investigative interest focused on the problem and few, if any, reports on the problem can be found in recent publications. I regard most of the references cited here and listed in the bibliography as key works that constitute only a small sample of the earlier work that is now over half a century old. The interested reader, however, will find it well worth his time to include these important references in his material.

A serious consideration of the problem at the present time becomes a search for correct concepts rather than additional data. With so little understood in the face of mounting research data, finding the proper approach and the correct perspective becomes the first concern. At the present time there are actually no concepts of menstruation to evaluate. The phenomenon is a puzzle to clinicians, and despite intensive scientific research of an entire century, not so much as a theoretical explanation has been the result. No one seems to have the foggiest idea of why women bleed monthly during their reproductive years or what possible connection the bleeding has to reproduction.

On the one hand, the "clockwork" regularity and the universality of monthly bleeding in women during their reproductive years are taken as evidence by clinicians that menstruation should be regarded as a normal physiologic process. Accordingly, the terms "functional uterine bleeding" and "dysfunctional uterine bleeding" have been devised by those clinicians keen on defining the phenomenon, with its variations and aberrations. On the other hand, such assignment of a physiologic role to the bleeding does not explain the phenomenon and is neither enlightening nor acceptable. Bleeding characterizes tissue trauma, and indeed the endometrium is regularly injured and becomes ulcerated before it heals and bleeds during menstruation. The endometrial ulcer and its healing are indicative of a pathologic state, leaving clinicians perplexed and researchers confused and far from the correct concept of menstruation.

Although other mammals closely related to humans also menstruate, they, along with humans, represent only the few existing mammalian species that menstruate. It is not a coincidence that these mammals have developed the largest brains. In most mammals estrous cycles occur without any endometrial hemorrhage. The significance of menstruation is its role in reproduction of those mammalian species that do menstruate in their estrous cycles. It should be noted that the menstruating mammals in existence today are not distantly related species scattered throughout the various mammalian orders, but are closely related forms that are concentrated within a single primate suborder, the Anthropoidea. This indicates the evolutionary basis of menstruation. Of the two infraorders composing Anthropoidea, the Catarrhini, which includes the human, bleed visibly during menstruation, while the Platyrrhini bleed only microscopically or not at all. The Anthropoidea, most of which visibly menstruate, are referred to in the following chapters as "upper primates" in contrast to the other primate suborder, the Prosimii, which, like infraprimate mammals, do not menstruate in their estrous cycles. Reports from more recent field research are based on observational and descriptive methods in which only visible menstrual bleeding can be detected and reported.

Also, reproductive cycles in mammals studied in their wild state become altered in the same animals subjected to laboratory studies in captivity. The intense laboratory studies of mammalian reproduction over half a century ago, like current field studies, were based on observational and descriptive methods that revealed the gradual stepwise progression in reproductive structure and function from one group of mammals to the next more advanced group. Older studies concentrated on structural or anatomical changes, while more recent field studies emphasize behavioral changes. Both approaches confirm gradual evolutionary change in behavior, function, and structure. Although reproductive endocrinology research may reveal a marked difference in endocrine physiology between related forms, results confirm gradual phylogenetic progression in mammals. Menstruation makes a gradual appearance in the evolutionary course of upper primates. The reproductive cycle characterizing most mammals, i.e., the estrous cycle of lower primates and infraprimate mammalian orders, undergoes little change as bleeding becomes added to the endometrial healing process in upper primates.

Efforts to explain the addition of bleeding to endometrial healing in mammalian reproduction bring into question why endometrial injury should occur in the reproduction of mammals. To persist in focusing on direct efforts to explain the bleeding on the basis of endocrine physiology alone, with neglect of the larger question, reduces such studies of menstruation to futility.

The proper approach to clarifying menstruation must be an orderly one in which reasonable answers to larger questions are established first—answers that are sufficiently adequate to provide the framework and foundation to include lesser questions. The pivotal question of why the endometrial membrane is damaged in mammalian reproduction opens the still larger unanswered question of what characterizes mammalian reproduction. What alterations in terrestrial vertebrate reproductive function and structure have accounted for the emergence of reproducing mammals from reproducing reptiles? Explanation of the evolutionary origin of menstruation lies therein, but menstruation is only one of several mammalian reproductive phenomena with an evolutionary basis that is yet to be explained. The formation of an ischemic ulcer in the endometrial surface of humans, as well as dysmenorrhea, is an example of such unexplained reproductive phenomena with origins that can be traced back to transition of reptilian reproductive physiology into mammalian reproduction.

Reptilian characteristics can be discerned in reproductive structure, function, and behavior of the most primitive mammalian forms in existence today, the Prototheria, a subclass of egg-laying mammals comprising Monotremata. Less reptilian and more mammalian specialization can be seen in reproductive cycles of marsupials comprising the subclass Metatheria. Still fewer reptilian characteristics are found in the more numerous orders of the subclass Eutheria, which are the placental mammals, the dominant group of existing mammals. Although interesting and important phylogenetic modifications have occurred in the cloaca and distal organs of the reproductive systems in these mammalian forms, it is the phylogenetic change in the three proximally located reproductive organs —ovary, oviduct, and uterus—that accounts for menstruation in primates.

The ovary is by far the oldest of the three organs, and its role in reproduction of plants as well as animals is clear. It produces the female gametes, the egg cells, endowed with sufficient stored nourishment and energy to support the zygote through its segmentation stages to larval formation.

The oviduct is a much younger and more recently formed organ than the ovary but still far older than the uterus to which it gives origin. The phylogenetic course of the oviduct is irregular in its adaptive modifications in function and structure. It began as an abdominal pore in our ancient ancestral, aquatic vertebrate in which the ovary dispensed its eggs into the peritoneal cavity and thence through the pore to the outside for external fertilization, development, and growth. In vertebrates that adapted to terrestrial environments, the internal and external openings of the pore lengthened into a funnel-shaped tube to transfer the eggs to a specific, moist, external location for fertilization and development. As vertebrates adapted further to increasingly dryer niches, behavioral changes gave origin to coitus and internal fertilization. Muellerian tubes then acquired the additional function of transporting sperm in the opposite direction, inward and proximally to the egg, before transporting the egg externally.

As vertebrates continued to progress in their adaptation to terrestrial existence, transport of the egg to the exterior during reproduction became more specialized, and specialization of structure followed in the distal portion of the tubes. A whole new organ, the uterus, evolved. Although irregular in its phylogenetic course, the role of the oviduct, like that of the ovary, is clearly understood. In contrast, the role of the uterus in mammalian reproduction remains far from clear. Clinicians must contend not only with its bleeding performance in humans, but also with the accompanying dysmenorrheal pain and cramps and the necrosing ischemia of the endometrial surface. These mysterious phenomena emanating from specialization of performance in the human organ underscore the inadequacy of our understanding of uterine physiology and the role of that organ in reproduction.

Gradual evolutional change in reproductive functions begin first as behavioral change, followed by physiologic changes that finally result in structural change. The reader interested in how genetic change accounts for new function and structure must turn to works in the field of genetics. Here, the broader changes that did take place over long periods of time in response to natural selection pressure take priority over establishing the minute steps in their progression.

Although other vertebrate classes exhibit modified structure and function in distal Muellerian tissue that may resemble and carry out some of the functions of the mammalian organ, nevertheless only mammals possess a uterus and uterine physiology. Advancement of uterine physiology appears as a prerequisite to the rise of mammals.

Following the origin of the uterus, a still newer organ, the mammalian placenta, developed. A similar modification of egg membrane physiology and structure can be found in some placental reptiles, but all steps in the development of this organ in mammals can be seen in the spectrum of existing mammals despite the dearth of aplacental forms. Although closely related to and dependent upon uterine physiology, the placenta arose after uterine function and mammals were already in existence. Placental physiology and structure in mammals appears to have been an evolutionary advancement of uterine function and structure.

The fossil record reveals that existing egg-laying mammals, the monotremes, are offshoots of a marsupial-type ancestor, an example of retrograde evolution, which in this case was most likely in a mammalian form possessing a primitive uterus. Therefore, egg laying in Monotremata cannot be considered a step in the evolution of mammals, but rather an offshoot of the mammalian lineage. The fossil record also reveals that marsupials were once the dominant form of mammal before placental mammals began to replace them. The fossil record does not and cannot reveal (as vividly as the existing mammalian forms reveal the origin of the placenta) all the soft tissue changes that resulted in gradual development of the uterus during emergence of mammals. The earliest uteri and uterine function had already appeared before mammalian placentae developed from egg membranes. The portion of the mammalian evolutionary path that is continued so vividly in existing mammals is but the most recently developed "tip end" of the path. This clearly visible segment of uterine evolution that includes only the latest portion of the path also includes that part of uterine evolution that accompanies mammalian placenta origin and development. The actual origin of the uterus, unavailable in the fossil record, is beyond study except for the "hints" remaining in the ontological stages of its surviving lineages.

Although they are only hints, nevertheless they are hints that reveal those evolutionary steps through which uterine function and structure arose in mammalian reproduction and thus provide the basis for correctly explaining puzzling phenomena in the human organ, including menstruation.

Part One
Biologic Roots of the Human Reproductive Cycle

I
Origin of the Estrous Cycle

To understand menstruation the menstrual cycle must be recognized as a special form of the estrous cycle. Although some clinicians may regard the two as separate entities, any apparent difference between the two is not supported by evidence. The menstrual cycle is an evolutionary modification of the mammalian estrous cycle in upper primates. Lower-ranking primates in their reproductive cycles without menstruation display the estrous cycle that is typical in infraprimate mammalian reproduction. A sufficient number of living primates have already been studied adequately enough to provide the intervening stages of transition from the typical mammalian estrous cycle to the menstrual cycle of the human. The menstrual cycle, therefore, having evolved from the estrous cycle, is too deeply rooted in estrous cycle physiology to be defined as a separate category of functions.

Reproduction of primates in their natural habitat is not always easy to study. By far the most studied primate is the human. However, the human menstrual cycle represents a very late and recent evolutionary modification of the estrous cycle in mammals, and indications are that it probably represents the most evolutionarily advanced form of the cycle. Although menstrual bleeding may be the most striking development within the estrous cycle in upper primates, other less obvious modifications of the cycle have also developed. These other modifications involve physiological changes associated with behavior as well as anatomical change in the internal reproductive organs. Behavioral studies of primates in their natural habitat have revealed estrus, or the period of desire, to be flexible in the timing of female receptive behavior within the cycle. In some primates the period of estrus lasts longer than the sharply distinct periods of estrus found in infraprimate mammals. Primatologists have thus redefined *estrus* to mean libidinous behavior of the female even when she is not approaching ovulation. Still, all primates "retain a tendency to concentrate matings at mid cycle" (Hrdy and Whitten in Smuts et al., 1987).

Such variation in sexual behavior does not detract from the basic fact that estrus in the female is critical in bringing about the coitus that effects egg fertilization in mammals. Studies indicate that in both estrus and rut the central nervous system, which controls the organism's behavior, is a target tissue of reproductive endocrines along with the reproductive organs. The same endocrines that induce behavioral changes in lower mammals also induce behavioral change and menstruation in upper primates. There are no sharp differences in sexual behavior between primates that menstruate and primates that do not. Only the presence of cyclic bleeding determines whether the cycles are menstrual cycles or estrous cycles, and menstruation in some primates is not obvious gross bleeding but faint microscopic extrusion of a few red blood cells into the vagina (Goodman and Wislocki, 1935). Although transitional stages between estrous cycle and menstrual cycle have been subjected to intense descriptive studies, the reason bleeding has developed in the estrous cycle of upper primates, thus converting it into a menstrual cycle, has not been elucidated. Although the same endocrines—identical chemical substances—produce different tissue and organ responses in different mammalian forms, the differences—behavioral, physiological, and anatomical—are all evolutional. The evolutional path of behavioral change is separate and different from the evolutional path that produces bleeding of menstruation, even though identical endocrines may produce both. Behavioral changes within the cycle only accompany physiologic changes that result in bleeding.

Without future species present for comparison we cannot fully interpret these changes, but the relatively short, intense periods of estrus seen in lower mammals appear to be changing to less intense but longer periods of desire within the primate cycle. The estrous cycle is gradually changing, but it is not yet completely replaced by a new and different form of cycle. We may not know exactly what form the estrous cycle is changing into, but estrus in the primate cycle is not disappearing as some might be inclined to think. The human menstrual cycle still includes estrus, or desire, that is modified and changed and must be recognized as evolutionally changing rather than disappearing simply because its intensity appears less concentrated than that seen in lower mammals.

Although we may not understand these evolutional modifications at first and simply describe them as characteristics of the upper primate estrous cycle, nevertheless their significance is much greater than a simple description may reveal. They represent the evolutional changes in mammalian reproductive functions, specifically, changes in placentation that have created and produced the upper primate. Evolutional changes from Ungulata to Carnivora and from Carnivora to Insectivora as well as from Insectivora to lower primates are well recognized in the structure, function, behavior, and adaptation in the adults of mammalian orders. These evolutional changes can be correlated with the type of placenta that produces each mammalian order. The order of aplacental mammals, Marsupialia, shows variations in body morphology and behavior, but the various groups comprising the order do not show the significantly advancing levels of central nervous system development, which is seen in placental mammals.

In lower placental mammals, the egg membranes greatly expand their respiratory surface late in gestation and support the metabolism of a larger-brained fetus. This path is greatly advanced in primates that menstruate in their estrous cycles. The evolutionary changes in nonplacental mammals, which incubate small uterine eggs, are accompanied by relatively minor variations and modifications in the estrous cycle. The same is true of infraprimate placental mammals in which early egg nutrition is also from yolk and uterine gland secretion. The evolutional changes that previously advanced certain lower-primate branches to their presently higher-primate status, with cyclic bleeding in their estrous cycles, involve evolutionary changes in early egg nutrition.

Although the adult mammalian forms that comprise the various orders of placental mammals may have been shaped by natural selection, the developmental level of their central nervous systems, rather than body morphology, determines their rank in the mammalian hierarchy. Advancing development of their central nervous system rather than their adult mammalian form is a product of advancing placentation physiology, i.e., the development of a better, more efficient placental exchange within the egg membranes. Evolutional advances in placental structure and function occurred first in the late gestational stages of placental growth as the egg membranes attained their mature limits. This is the beginning of evolutional change that actually accounts for the various orders in lower-ranking placental mammals. Thus the expanded placental respiratory surface of the egg late in gestation is far more significant than simple anatomical descriptions of egg-membrane development would indicate, but the expanded exchange surface late in gestation did not evolutionally alter early, small-egg surface and estrous cycle physiology. Only when advancing evolutional improvement in egg respiration (placentation) progressed to the earliest stages of egg nutrition in the creation of upper primates did the estrous cycle become evolutionally altered to produce menstrual bleeding.

Estrus is cyclic in mammals and birds, classes that, like Reptilia from which they arose, are adapting terrestrial vertebrates. Birds, the most recent offshoot from reptiles, have continued to reproduce in the reptilian pattern of laying large, shelled eggs that hatch by their own physiology; only relatively minor evolutionary modification of this basic reproductive pattern has produced the bird lineage. In contrast, the earlier offshoot from reptiles, the mammals, arose through marked evolutionary changes in the reptilian reproductive pattern. These evolutionary changes must be regarded as revolutionary, as they have altered the entire reptilian reproductive cycle and account for the origin of the estrous cycle in mammals. Because the menstrual cycle is an evolutionary advancement of the mammalian estrous cycle, the origin of the estrous cycle in mammals must be considered by those seeking to understand its change to the menstrual cycle in upper primates. Here, reproductive cycles of reptiles, birds, and mammals must be further compared with special reference to human reproductive physiology.

Although reptiles adapted through development of a large egg, the reptilian branch that gave rise to mammals consisted of small reptilian forms that laid small eggs. Seemingly, the gross size of the ovarian egg was on a retrograde path of evolution in the mammalian lineage, beginning a reduction in size that, in existing placental mammals, has diminished to a diameter of only a fraction of a millimeter. This evolutionary reduction in ovarian egg size was the first and most prominent change and one that has accounted for all subsequent evolutionary change in the mammalian lineage. In the beginning it was the only component of the evolutionary path. It involved only the anatomical size of the egg when it was discharged from the ovary. Later components of the path include opposite change in size of the tubal egg. The initial size of the tubal egg was, of course, the same as its ovarian size. However, in later phases of development during the reproductive cycle while incubated and nourished within the tube, the egg in the mammalian lineage continued its evolutionary development through enlargement.

With egg development in placental mammals freed from the limited source of stored yolk, enlargement has been carried to comparatively enormous sizes over longer periods of gestation in which the egg hatches the larger fetus rather than the smaller embryo. In aplacental mammals, the egg is hatched after a brief gestation period lasting only a few days and delivered while it contains only an embryo. In placental mammals, however, the egg is hatched and delivered after a gestation lasting several months, a time in which it enlarges way beyond the size of eggs in other classes of terrestrial vertebrates, to produce a fetus with a larger brain than an embryo.

The enlargement of the tubal egg in the distal portion of the oviduct was not an accompaniment of diminishing ovarian egg size, nor was it simply part of the same evolutional change. It was on the same evolutional path but in a more advanced step that followed after the initial step of reduction of ovarian egg size. After reaching maximum diminution in anatomical size in aplacental mammals, the ovarian egg in placental mammals has continued the evolutional path of size reduction functionally. Tubal, or rather early uterine, egg enlargement began after the ovarian egg lost most of its yolk. Yolk is a more limited source of oxygen and respiratory surface than maternal tissue surface, and a large yolk mass beginning to diminish was not a source for evolutional enlargement of respiratory surface. The smallest ovarian eggs in mammals, the ovarian eggs of placental mammals, contain only the amount of yolk necessary to sustain the fertilized egg through implantation or fixation to the endometrium. Anatomical reduction of ovarian egg size was simply the evolutional prerequisite to functions that enlarge the egg in the tube. Since egg yolk is so limited as a source of oxygen for early embryogenesis, only a small-brained embryo develops from a yolk source. For the large-brained embryo and fetal formation that subsequently evolved in placental mammals, large-yolked eggs reduced the size of their huge yolk masses as their chorions acquired more respiratory surface and respiratory exchange with expanding tubal endometrium. Yolk and its enclosure, by developing embryonic gut structure, became retired as the egg evolutionarily lost yolk, i.e., nutrition of the egg switched from yolk and entoderm to chorionic ectoderm and endometrial surface. Thus the size of the egg reduced first, and only embryos were hatched from the early "tubal uterus." This still occurs in presently surviving Monotremata and Marsupialia. The tiny eggs of placental mammals, however, expand their chorio-endometrial respiratory surfaces enormously, at first delaying embryogenesis until adequate respiration for a larger-brained embryo becomes established. These evolutional stages are considered in more detail in the next chapter.

Anatomical reduction of the ovarian egg size resulted in the tremendous changes that characterized the mammalian reproductive physiology that followed. Some examples of these changes to be considered are: (1) removal of the yolk-mass barrier that freed the egg to develop larger respiratory surface, (2) egg incubation internally as well as externally with control of higher fetal body temperature, (3) development of two new organs, the uterus and placenta, (4) greater maternal membrane surface, (5) greater fetal oxygen consumption, (6) development of larger fetal brain and central nervous system, (7) mechanical support of a large egg during its incubation, (8) miniaturization of structure and function in the initial phases of the reproductive cycle, (9) enlargement of structure and function in the later stages of the cycle, (10) conversion of tiny egg structure to endocrine glands, and (11) mechanical hatching internally with delivery of the huge egg. All of these changes in mammalian reproductive structure and physiology are closely related and gradually progressed together after the ovarian egg acquired its maximum reduction in anatomical size. Together they form the gradual steps in the advancing evolutionary path of mammals in which ovoviviparity converted to viviparity.

This is not to imply that only improved physiology of placental exchange accounts for all evolutionary change in mammalian body form by fostering anatomical brain enlargement; the natural environment also has a direct influence in shaping evolutionary change in both external body morphology and behavior in the adult form. Brain and central nervous system enlargement does account for some change in body morphology through its anatomical size as well as through its physiological influence in advancing behavior, but influence of the natural environment ultimately shapes all evolutionary change. The natural environment even accounts for the evolutionary changes in placentation. It was adaptation to the natural environment that previously brought on development of the large-yolked egg in land vertebrates. Reduction in egg size and evolutionary advancement of placental function and structure, however, are the most immediate and major causes of physiological change in mammalian reproduction that account for progressive advancement and enlargement of central nervous systems in the various orders of placental mammals.

One of the first changes in the reproductive cycle that resulted when the ovarian egg became smaller was the replenishment of its yolk by absorbed oviductal (primitive uterine) secretion. Initially, reptilian egg size and embryonic size at the time the egg was ready to hatch were maintained essentially at a constant volume, the volume the egg has when it hatches externally. The hatchling thus remained small in size, no larger than the volume of its shelled enclosure. This pattern is certainly evident in existing aplacental mammals. In reptilian style, embryogenesis, utilizing stored egg yolk, begins and proceeds while the egg is moved toward the uterine modification of the tube. Upon reaching the glandular, or uterine portion, of the tube, the egg absorbs additional nourishment in the form of uterine secretion through its chorion into its embryonic yolk sac, where yolk and secretions are mixed and remain stored in the egg for embryonic absorption. As the ovarian endowment of stored yolk diminished in the egg of these past transitional forms, embryogenesis in the egg became delayed, as prominent egg ectoderm and entoderm development necessarily preceded embryogenesis, to absorb and store additional uterine nourishment for subsequent embryonic use.

Oviductal function controlling movement of the egg through the oviduct and secretions and functions of the uterine portion of the tube continued to remain under the influence and control of the ovary. Specifically, within the ovary the residual egg bed following ovulation began to develop into an endocrine gland (corpus luteum) that continued ovarian control of tubal function. The eggs, previously very specialized in structure (for external hatching) and later becoming modified in function and structure, were produced sequentially not simultaneously and passed into the tube individually as they each became coated with albumin water. The shell glands located distal to the early forming uterus gradually became vestigial and disappeared in placental mammals with tiny eggs. The number of eggs produced at ovulation were fewer than the number produced in the reptile clutch, and the total egg mass probably did not exceed a number of eggs greater than the number the uterine portion of the oviduct could accommodate. As higher body temperature developed, body folds and pouches continued incubating the embryo after it hatched, and modified sweat glands of the parental skin provided water and food for further embryonic growth. These earlier mammalian reproductive functions are exemplified in existing aplacental mammals in which the mammalian reproductive cycle can be seen to be immediately under ovarian endocrine control.

The various species of living primates, which demonstrate the latest of these evolutionary changes, indicate the problem of interpreting gradual evolutionary change in existing forms. Variation in living species can be confusing in attempting to determine an evolutionary path. Even with all stages of an evolutionary change represented and spread out in full view, we may not necessarily understand them. The change and the length of the time span over which the change occurred may appear at first to be very long; yet the time span still may not be sufficient to reveal the evolutionary direction in the face of so much variation and divergence of lineages. In searching for the means to interpret correctly these evolutionary changes in mammalian reproduction, a broader perspective that covers greater evolutionary change over a still longer span of time can be more revealing. Over longer spans of time than that represented by existing mammals, "dead branches" of the "evolutionary tree," which may be confusing, have been culled by natural selection, so that the main branch or trunk—the evolutionary path—may be determined more clearly. A larger evolutionary time segment, extending far enough into the past to include the origin of the estrous cycle, may be looked to for understanding the direction and course mammalian reproduction has followed in forming the menstrual cycle in the primate order.

It is not enough to simply establish the steps or stages in the transition of reptilian reproduction to mammalian reproduction. Documenting evolutionary steps even as detailed as those we can discern in transition from estrous cycle to menstrual cycle does not necessarily leave us with an understanding of the changes. Comprehending the whole evolutionary change resulting from natural selection, however, is paramount for a clear understanding.

The estrous cycle is well known from descriptive studies, but precisely what it is has not been made clear. Although the estrous cycle in female mammals evolved from estrus and reproductive functions of reptilian female ancestors, reptiles were a much more extensive group of vertebrates than the first mammals. With a body temperature that tends to vary with the environment, reptiles have flourished in tropical areas. Their reproduction appears to be more seasonal than cyclic, more dependent upon environmental temperature change, and is not as independent, consistent, and fixed in behavior and physiology as mammalian estrous reproduction.

The reptilian equivalent of the single mammalian event, i.e., the estrus that brings about coitus and fertilization is actually two reproductive events timed differently in the season. Estrus in female reptiles is an event of receptivity for coitus with the male. The sperm is stored in the female reproductive organs because of lower reptilian body temperature and may fertilize eggs over several succeeding reproductive cycles and subsequent breeding seasons. The actual reproductive cycle in female reptiles, however, is deposition of fertilized eggs that hatch on their own after incubation at environmental temperatures. Evolutionary merging of estrus and coitus with egg production in the female became necessary in the evolution of mammalian forms developing high, regulated, and relatively consistent body temperature—too high for stored sperm to survive long in the female body. The higher body temperature of the emerging mammal made fresh or unstored sperm necessary for egg fertilization. Estrus and coitus had to be timed simultaneous with the production of the egg (ovulation). The reptilian reproductive cycle, in which large, specialized ovarian and tubal eggs were extruded for outside incubation and hatching at a later time, changed in the rising mammalian lineage. It became a cycle in which smaller and eventually tiny, unspecialized ovarian and tubal eggs were retained, incubated, and nourished in the oviducts. The egg, incubated internally in the mammal, thus became completely protected from vicissitudes of environmental temperature.

The typical reptilian reproductive cycle ends with deposition or oviposition and abandonment of the eggs. In the mammalian and bird lineages, merging the reptilian estrous cycle with the reproductive cycle to ensure fertilization necessarily resulted in estrus occurring in the early stages of the reproductive cycle as the egg was being discharged from the ovary.

The change from a reproducing reptile to a reproducing mammal took place in the comparatively dry, oxygen-rich environment of terrestrial niches. The mammal, with its larger brain and higher body temperature, is a much greater oxygen consumer than the reptile, just as the reptile is a greater oxygen consumer than its amphibian ancestors, which produced even less body heat. Progressively greater oxygen consumption, body heat production, and body water preservation in metabolism of each succeeding class of terrestrial vertebrates represented advancing adaptation to the dry terrestrial environment.

The first amphibians from which reptiles arose were not divorced from a water source and absorbed oxygen from skin and gills while in the aqueous environment, but when on land oxygen was absorbed through their moist membrane surfaces such as integument, pharynx, swim bladder, or lung. Body water loss through evaporation during short periods on land was replenished by frequent returns to the readily available water environment. Body heat gain and loss were more in equilibrium with environmental temperature than succeeding lineages that utilized more oxygen in their metabolism and produced more body heat as they progressed toward the reptile state. Acquisition of body scales confined all oxygen and water absorption to exposed body channels and their openings—mouth, pharynx, gastrointestinal tract, trachea, and lungs. The reptile had to drink for body water intake and had to breathe for oxygen absorption. Trachea and lungs functioned somewhat to preserve moisture and heat by acting as a heat exchanger, drawing atmospheric gases through tubes (tracheo-bronchial tree) and mixing them with discharging respiratory gases.

Scales greatly advanced body water preservation by reducing evaporation loss but were not too efficient in preserving body heat. Development of hair in mammals and feathers in birds, on the other hand, enabled preservation of both body water and body heat. Both these later classes of terrestrial vertebrates have adapted by utilizing more atmospheric oxygen in the production of higher, regulated, and more uniform body temperature. Besides a body covering of scales, as living forms show, the reptile also adapted in its reproduction by developing a very large egg, a virtual tank of food-yolk and water, supported and contained within an exoskeleton consisting of a porous calcium shell through which the egg breathes. This huge and very specialized egg breathes on its own and is laid in warm, protected areas such as areas of decaying vegetation, where it is abandoned to develop and hatch on its own. The egg contains all its requirements for developing an embryo, including food, water, volume, and respiratory surface. Oxygen and heat, and in some cases moisture, are absorbed from the environment. It hatches from within through physiology or behavior of the embryo, when embryonic volume is equal to and begins to exceed eggshell volume. The large egg of the reptile further advanced reptile adaptation to dry land by freeing it from the necessity of having to return to a water environment, to reproduce as its amphibian predecessors had to do.

The evolution of first mammals then birds came about through different evolutionary changes in reptilian reproductive physiology. The emergence of mammals resulted from severe change in the entire reptilian reproductive physiology, while birds emerged later from the comparatively unchanged reptilian pattern. What modifications did take place in emerging bird reproduction are relatively minor changes that took place in the later stages of the reproductive cycle. With gradual acquisition of higher body temperature, the egg was not abandoned but was incubated with adult body heat in the behavior of brooding. The bird embryo develops feathers rather than scales and begins to produce its own body heat in the late stages of its formation, leaving early embryogenesis in birds virtually identical to that of the reptile. The greater consumption of oxygen in metabolism and generation of body heat in the bird embryo is provided by the more expanded respiratory surface of the allantoic egg membrane. As in the reptile egg, the allantois and its surface enlarge as it stores embryonic metabolic waste as insoluble uric acid. This latter physiology not only preserves egg water but also provides an expanding respiratory surface as the embryo grows larger. All of these functions were altered and developed along a different course in reproductive physiology of the mammal.

In the first steps of the evolutionary reduction in the ovarian egg size, the typical reptilian reproductive pattern of function and structure persisted. All embryo nourishment was derived from stored yolk and water in an externally laid egg. Because of the reducing egg size, the hatchling probably was no larger in size than the embryo typically produced at parturition in existing aplacental mammals. But the small hatchling from the abandoned reptile egg possessed its scales and was immediately equipped for and capable of fending for itself in the environment of the time. True to the reptilian pattern, the hatchling was essentially a miniature adult that completed all embryonic development within the egg and simply grew in size after hatching, in some cases (dinosaurs), to an enormous size. In the evolving mammalian and bird lineages the higher developing body temperature of the hatchling was accompanied by additional evolutionary changes in parental behavior other than those resulting from merging of estrus and reproductive cycles. Not only was the egg not abandoned but it was brooded, and the hatchling was also brooded and carefully attended; it was too small to maintain its own body heat, exist, and fend for itself. Food, water, and body heat were derived from the parent in both mammalian and bird lineages. In the mammalian lineage the hatched embryo was supplied with body heat, sweat (water), and food secretions from the maternal skin, while in the bird lineage food and water were provided by parental feeding behavior. The small embryo-sized hatchling in both early mammal and bird lineages further developed its central nervous system only slightly as it grew after hatching. These evolutionary steps can be seen in existing orders of aplacental mammals as well as birds (see chap. 2). Primitive mammals in existence today, like birds, have small brains in comparison to recent placental mammals.

Among the first behavioral changes that probably occurred in the mammalian branch of reptiles was the ability to retain the small eggs inside the abdomen before hatching. Some species of present-day reptiles may deposit eggs for hatching (oviparity) where they occupy areas in which the weather is warm, but the same species may retain their eggs inside the body where they occupy areas in which the weather is colder.

Past forms in the early mammalian lineage also began to retain their eggs for variable lengths of time at first then eventually until the eggs hatched. Their eggs were adapted for hatching externally, and retaining them inside a reptile body (within the oviducts) until they hatched (ovoviviparity) as the means of protecting them from outside temperature changes was not incubation initially. The reptile body temperature was not sufficiently high to actually warm the egg. The reptile body served more to insulate and preserve that heat which the egg had previously captured or whatever heat the tiny embryo was able to produce, if any. However, this behavior and physiology was the beginning of internal or uterine incubation, later continued in more mammal-like forms with embryos and fetuses that did use more oxygen and produce more body heat.

Although the reproductive cycle in reptiles ends with egg laying and abandonment of the eggs, in mammalian and bird lineages the cycle ends later, after all feeding and attending behavior to the hatchling is accomplished. In birds, removal of eggs from the nest may induce the female to lay another clutch, just as removal of the entire litter of newborns in mammals tends to induce estrus and another reproductive cycle. The large bird eggs are visible and palpable to the brooding female, and their absence provides the stimulus to the hypothalmic-pituitary-ovarian axis to initiate a new reproductive cycle. On the other hand, in the mammalian lineage, with progressive reduction in ovarian egg size, the small egg became undetectable either visually or palpably. Its evolutionary disappearance early in the reproductive cycle triggered or stimulated the neuroendocrine system to initiate a new cycle through a stimulus other than through the organs of special sense. Experimental removal of all tiny mammalian eggs early in a specific cycle until recently has not been practical in placental mammals, but in the infertile estrous cycle the unfertilized eggs become effectively removed by death and autolysis, providing the stimulus that induces a new cycle. Also, removal of the embryos from the teats in the pouch of marsupials later in the reproductive cycle will induce the estrus of a new reproductive cycle within a few weeks (Hartman, 1923). In the human, menstrual cycles of the nursing female tend to be suppressed but return sooner when the nursing infant is removed.

The later phase of reptilian egg hatching that has persisted with much less change in the bird lineage underwent much greater modification as it evolved into the late phase of the mammalian cycle. Evolutionary change in the mammalian egg late in the cycle induced severe changes in maternal anatomy, physiology, and behavior. Already estrus and coitus had greatly altered the initial phases of the mammalian cycle. These marked alterations in the late phases now resulted in complete change of the entire cycle. The rigid eggshell that was burst by the hatching embryo in externally hatched eggs was evolutionarily retired when the ovarian egg became internally incubated and enlarged through absorption of larger quantities of uterine nourishment. Indeed, the eggs of existing aplacental mammals, varying in size from 4 to 5 millimeters in monotremes to less than 0.32 millimeters in marsupials possess a shell membrane, and vestiges of shell glands are found in the reproductive organs (Hill, 1911).

As the mammalian egg became incubated and hatched internally, new and different egg-hatching physiology developed within the incubating tissues of the distal portion of the oviduct, gradually transforming that tissue into a uterus. The new organ performed the mechanical egg hatching that the mature embryo did in the oviparous state. This modified mechanical function of the distal oviduct began as nutrition for the developing embryo shifted from stored egg yolk to tubal and uterine secretions, and oviparity became ovoviviparity. Retaining the egg in this distal portion of the oviduct until it hatched progressed to actually mechanically opening and hatching the egg during its expulsion from the maternal body. Delivery of the egg and hatching of the egg were merged. These new tubal functions, i.e., first uterine functions, were simply modifications of egg-moving functions of the oviduct. First, the eggs were delayed and stored in the enlarged channel at the distal end of the tube to absorb nourishment. After a delay the swollen eggs with embryos were squeezed into the birth canal, bursting their membranes and freeing the embryos during their expulsion.

Initially there may have been as many eggs as we see today in aplacental mammals, and each egg, as in marsupials, was probably mechanically ruptured and hatched during its expulsion by the peristaltic action of the tubular uterus. The appearance of the uterus for incubation and nourishment may have been an evolutionary constraint to the number of eggs produced in the ovary during each cycle. Functions within the longer-incubated egg began to take over ovarian control of the uterus and trigger uterine hatching functions. Hatching or direct mechanical opening of the egg from inside the egg through shell-piercing behavior of the embryo, in the manner reptile and bird embryos hatch, was replaced when the newly modified distal tube acquired emptying functions of a uterus. The stimulus triggering the primitive egg-opening functions in the uterus of aplacental mammals comes from endocrines of the ovary that control the functions of Muellerian tissue, while the same endocrines from a further advanced embryo and its egg membranes trigger the uterine hatching functions in placental mammals. The mammalian egg, in addition to inducing the formation of its own nutritional bed in the uterus, induces its own hatching functions in the uterus both directly and indirectly in assuming its complete control over uterine delivery mechanics. Its direct control is its morphological (physical) presence and enlargement within the uterus. Its indirect control is through the blood-borne endocrines it discharges into the maternal circulation. Because the egg of higher placental mammals grows in much closer association with uterine tissues, its discharge of endocrines from within uterine tissue also directly affects uterine tissues in advancing uterine egg-hatching and delivery functions.

While the early emerging mammalian forms that went through these evolutionary steps produced no surviving offshoots that could now be studied in detail, nevertheless these evolutionary steps did occur. They must be recognized as having occurred even though they are almost entirely lost in the evolutionary modifications of surviving lineages. So far, the fossil record has provided a glimpse only of fossil embryos of ancient marine reptiles, but the question of their oviparity versus viviparity is still ambiguous (Sanders, 1988). Numerous small, soft embryos within the fossilized abdomen after the abdominal wall deteriorated would be rare, if at all possible. Were such available that demonstrated soft tissue reproductive functions, perhaps some of the mystery of the origin of mammalian reproductive physiology could be dispelled.

Egg development in reptiles, birds, and mammals changed only to the extent they were actually driven to change, by the progressively diminishing ovarian egg size. As long as the egg continued to be large because of the large amount of stored yolk, the embryo was evolutionarily committed in its physiology to utilize that source through absorption from the inner surface of its membranes that encapsulated, and were in contact with, the stored source. As the stored source (of food, water, volume, and oxygen) diminished and was used up sooner in progressively smaller ovarian eggs that were internally incubated, the embryonic membranes gradually began to switch to their outer surfaces in contact with the sources provided in surrounding maternal tissues. These sources proved to be more than adequate for the developing egg's necessities. Maternal tissues provided more than simply larger amounts of food, water, volume, oxygen, and heat. Compared to the external environment, the environment in which the reptile and bird egg develops, maternal tissues provides a virtual unlimited source of all the egg's needs.

The mammalian egg reached the limits of this new environment in its evolution of primates. Offshoots from the advancing evolutionary steps account for past and present orders of mammals. Of all the egg's necessities, the most limited in maternal tissue was oxygen, which, though available, was less than the atmospheric oxygen content of 20 percent. Steps in acquiring additional egg respiration established the evolutionary offshoots that comprise existing orders of placental mammals. The tiny egg is nourished during a lengthy incubation period until, without a supporting calcium shell, it attains a huge size and weight that, in the human, can be as much as 10 percent of the mother's weight. Another 2 percent or so of the human mother's weight at term can be from growth and enlargement of the tissues that hatch (deliver) the egg, i.e., the uterus. Such enormous enlargement of the uterus is required in the mechanics of supporting, opening, and expelling such a huge membranous egg.

Progressive intimacy of the endometrium and the egg in procuring more nourishment for the latter evolutionarily invoked development of stronger delivery mechanics in the uterus. Mechanical egg-opening and delivery functions (parturition) in mammals necessarily became dominant over egg-nourishing functions in regulating and controlling the reproductive cycle. These functions operate to open and deliver or hatch the egg when the egg is delivered or expelled at any and all of the stages of its development. Although the mammalian egg changes in its internal morphology during gestation, the mechanics of its opening and hatching physiology in the uterus do not. This fundamental physiological pattern of mechanical function and structure of the uterus, i.e., egg-hatching and delivery functions that accompany evolutionary changes in egg nutrition during mammalian gestation, only grows and enlarges along with the egg. As the egg enlarges, its delivery mechanism must necessarily enlarge also but without change in pattern. Delivery functions are triggered by egg physiology that induces the uterus, to whatever size egg and uterus have grown, to apply the same mechanical hatching and delivery pattern to the egg irrespective of the latter's development and size. Whether the egg is delivered in its initial tiny ovarian size, as it is in an estrous cycle, or is finally delivered in its huge, mature size at term, the uterus applies the same delivery and hatching mechanics.

The developing aplacental embryo before hatching exerts no influence whatsoever on the reproductive cycle, i.e., no mechanical, no endocrine, or any other effect. Even when the eggs are all unfertilized, sterile, and dead, the ovary carries the reproductive cycle of the uterus through its delivery and hatching mechanics of parturition; however, only uterine secretions and disintegrating and liquefied eggs are delivered. The only means of differentiating between an estrous cycle and a gestation cycle in aplacental mammals is to open the uterus and examine the fluid-suspended eggs for embryos. Hatched live embryos, however, do have an effect on the reproductive cycle; their suckling continues the reproductive cycle through its final lactation stage. On the other hand, the absence of nursing embryos leaves unstimulated the hypothalmic-pituitary path that sustains lactation. The resulting cessation of lactation function then terminates the reproductive cycle; an enlarged uterus continues to shrink or involute and lactation ceases. Another reproductive cycle is then initiated by the hypothalamic-pituitary system. Living, suckling embryos in the pouch thus continue the reproductive cycle, while death of all embryos in the litter (or eggs in the uterus) terminates the cycle.

Similarly in placental mammals lactation physiology and egg-hatching physiology necessarily overlap and are thus very closely associated. Shrinkage, or involution of the enlarged uterus after egg hatching, is not only maintained but is accentuated, if not induced, by lactation functions. When the egg, or embryo, dies before suckling and lactation occur, a new reproductive cycle is instituted sooner. If the human uterus and its blood vessels have not completed their involution, considerable uterine bleeding can occur in varying degrees at the beginning of the new cycle. These physiological relationships are extremely important to understanding menstruation. Lactation is rare in menstruating women who have never conceived, but monthly breast enlargement and nipple secretion are frequent. In all of this reproductive physiology up through existing aplacental mammals, it is still the ovary that immediately controls the reproductive cycle. Ovarian induction of estrous behavior in the central nervous system previously mentioned is essential in the early stages of the reproductive cycle followed by ovarian set-up of the mechanical egg-hatching function in the uterus. Actually, the ovary in mammals, in addition to establishing the nourishing egg bed on the endometrial surface, also induces estrus and establishes latent egg-hatching functions in the uterine musculature as it produces the egg. Follicle endocrines account for the responses of both the central nervous system and the entire uterus. These first two stages of the mammalian reproductive cycle that accompany egg production, i.e., estrus and egg-hatching functions, the ovary induces through two different groups of endocrines, the estrogens and the progestins. The estrogens from egg production in the ovary initiate the reproductive cycle by inducing uterine growth and estrus first. The subsequent addition of progestins establishes involuting capacity, the mechanical hatching and delivery functions. (Coitus, in some mammals, induces ovulation that results in progestin secretion in the ovary.) The third and last stage of the reproductive cycle, the lactation stage, is under the control of suckling embryos. When suckling embryos fail to appear, the hypothalmic-pituitary system closes the mammalian reproductive cycle by stopping its breast stimulation and its inhibition of ovarian function and initiates a new reproductive cycle by stimulating ovarian reproductive functions.

The most frequent cause of failure of suckling embryos to appear and function in the mammalian reproductive cycle is failure to accomplish egg fertilization. In this respect, adaptation of the mammal can be regarded as much more complex than the adaptation of either reptiles or birds. The environmental niche to which mammals have been adapting is not an environment in which male and female can always be brought together for instant coitus at the whim of a quick physiological requirement. Sterile, unfertilized eggs, of course, perish; the failed cycle is closed out, and a new one instituted while the breeding season is still on. With ovarian eggs so small at the beginning of the mammalian cycle, little tissue of the old, failed cycle has to be discarded before a new cycle can be initiated.

In more advanced placental mammals, with even better regulation of higher body temperatures than in aplacentals, reproductive cycles appear to be more protected from, and less responsive to, seasonal change in temperature. The breeding season and reproductive cycles tend to extend over longer periods of time and, in some species such as the human, throughout the year. Placental mammals, with all egg incubation completed in the uterus before hatching and delivery, have greatly extended the length of the reproductive cycle beyond the initial control and influence of the ovary. Sterile eggs from failed fertilization are not nourished and incubated in placental mammals as they are in the uterus of aplacental mammals. Unfertilized and dead placental eggs are discharged by means of the uterine hatching and delivery physiology induced by the ovary. The fertilized and living egg in placental mammals extends the length of the estrous cycle (reproductive cycle) beyond the phase of ovarian control by delaying the ovarian hatching and delivery functions, specifically the uterine emptying functions. It does this by first replacing the function, influence, and control exerted by the hypothalmic-pituitary endocrine system that controls the ovary. The incubating uterine egg thereby continues and advances the endocrine function of the ovary (maintains the function of the corpus luteum, which sustains the latent uterine emptying mechanism). The hatching and delivery functions that the ovary induces in the primitive uterus of aplacental mammals are thus only postponed and magnified during the lengthier gestation of placental mammals. As the uterine egg enlarges, it not only enlarges the uterine delivery function by enlarging the uterus, but it also completely replaces the endocrine functions of the ovary by supplying greater quantities of ovarian endocrines. The greatly enlarging gravid uterus of higher placental mammals outgrows the endocrine support of the small corpus luteum in the ovary. The steps of this endocrine switch from ovary to uterine egg can be seen in the existing orders of placental mammals. The embryo of placental mammals thus controls the reproductive cycle before hatching in contrast to the embryo in aplacental mammals that controls the cycle only after hatching.

This shift of embryo control from after hatching to before hatching, i.e., the progressive control of the mammalian reproductive cycle by the incubated egg and embryo before hatching and delivery, is a continuation of the original evolutionary path of ever-diminishing ovarian egg size that is carried to advanced stages in placental mammals.

That gestation in placental mammals extends beyond the length of the estrous cycle has been known for some time (Hartman, 1923; Amoroso, 1959). Any difference between the functions that bring the estrous cycle to an end and the functions that end mammalian gestation have not been elucidated. If such differences exist, then, as placental mammals evolved from aplacental forms, the estrus-ending physiology would have had to be retired and replaced with new and different gestation-ending physiology (parturition). There is no evidence that such evolution change in physiology either could or did take place. While the length of gestation in aplacental mammals is the length of the estrous cycle, the latter ends in parturition of embryos or delivery of degenerating unfertilized eggs. Not only is the termination physiology of the estrous cycle and parturition the same, but the beginning of mammalian gestation and the beginning of the estrous cycle are also identical in both aplacental and placental mammals. A beginning estrous cycle is the beginning of pregnancy in the mammal unless cut short by death of the egg. Traditionally, clinicians, for very practical reasons, have calculated the beginning of human gestation from the time of fertilization of the egg, or the "fruitful coitus." These are early events in the mammalian reproductive cycle but do not mark its physiological or biological beginning. Fertilization, the actual entry of the spermatozoon into the egg, produces no physiological effects whatsoever in parental physiology. In the human, parthenogenesis does not naturally occur; egg development only follows fertilization, which also commingles ancestry. In placental mammals physiological results that follow egg development resulting from fertilization are continuation, prolongation, and extension of the reproductive cycle, i.e., the lengthening of estrous cycle physiology into physiology of gestation.

The length of mammalian gestation could not be less than the length of the estrous cycle, and as gestation lengthened in the rise of placental mammals, parturition functions established at the beginning of the cycle were simply delayed.

Definition and concept of the estrous cycle must be reconsidered. The developing egg, completely abandoned in the reptile, first began to control maternal behavior in both mammal and bird offshoots, as nesting and brooding behavior in bird reproduction presently demonstrates. With the continued evolutionary reduction in ovarian egg size in the mammalian lineage, its accompanying miniaturization of maternal structure eventually extended beyond structural and anatomical reduction in size to reduction in reproductive functions. Endocrine physiology of minute tissues and small organs to control much larger and still enlarging organ masses necessarily became more prominent in the mammalian reproductive cycle. With miniaturization of ovarian egg structure came miniaturization of associated maternal structure and function early in the mammalian reproductive cycle. Maximum expansion and enlargement of maternal structure and function is reached only during later stages in the reproductive cycle, as the developing egg physically grows and enlarges. The estrous cycle, therefore, is a complete, miniature mammalian reproductive cycle. Parturition functions in miniature end the estrous cycle, but when enlarged end gestation as well. Gestation in placental mammals is an enlarged, expanded, and prolonged estrous cycle. The estrous cycle and the mammalian reproductive cycle must be recognized as one and the same.

Much of the evolutionary path of mammals lies in a gap of extinct lineages, which leaves the presently surviving mammalian forms isolated from their predecessors. This gap increases the difficulty in clarifying mysterious reproductive functions in the isolated but continuing mammalian lines. The origin of the path that leads to menstruation in upper primates falls within this gap. Nevertheless, understanding menstruation first requires a clear understanding of the estrous cycle, which implies a reasonably clear understanding of its evolutionary origin.

Summary

The menstrual cycle is a form of the estrous cycle. Estrus brings about coitus at the fertile time in the reproductive cycle of mammals. The behavior of estrus is produced by endocrines acting on the central nervous system, and these same endocrines, acting on reproductive organs in upper primates, produce menstruation. Although identical endocrine substances can produce different responses in different mammalian forms, the different physiological and behavioral responses are evolutional. The evolutional path of behavioral change is a different evolutional path from the one that produces menstruation, despite the common biochemical cause. *Estrus* and *rut* behavior in humans is evolutionally changing but is not disappearing.

Changes in the estrous cycle of upper primates that include bleeding as well as behavioral change are due to evolutional change in placentation. The form of placentation determines the extent and stage of development that the central nervous system reaches in the embryo, and this placentation accounts for the various orders of mammals.

Mammals rose through marked evolutional change in the reptilian reproductive cycle in contrast to the bird lineage, which branched off later with little change in the basic reproductive pattern. Two new reproductive organs evolved in the mammalian lineage: first, the uterus, then the placenta. Placental function is a further advanced stage of uterine function.

Reduction in ovarian egg size at the time of ovulation established the estrous cycle and began the mammalian lineage. Elimination of the egg-yolk mass removed the barrier to early egg-membrane expansion and greater oxygenation of the developing egg. Reduction in ovarian egg size was accompanied by miniaturization of maternal structure and development of endocrine control of reproductive functions by the developing egg. The result left initial phases of the reproductive or estrous cycle totally under endocrine control in all infraprimate mammalian orders, with estrus occurring early in the cycle accompanied by changes in endometrial structure rather than the egg structure.

Expansion of membranes in the internally incubated egg in late gestational stages increased respiratory exchange that fostered the larger embryonic brain development and higher body temperature in existing mammalian orders. The estrous cycle of Aplacentalia, which was also their gestation or reproductive cycle, became extended in Placentalia. The estrous cycle–ending uterine physiology was delayed until after incubation was completed, advancing estrus-ending uterine physiology to labor and delivery physiology in Placentalia.

As increasing respiratory expansion of egg surface involved earlier stages of egg development during its implantation and attachment to the endometrium, in Placentalia characteristics of the initial phase of their estrous cycles became altered. These evolutional changes completed evolutional alteration of the entire mammalian reproductive cycle, resulting in the production of primates and the menstrual cycle.

In upper primates respiratory nutrition of the egg increased to the point of chorionic invasion into the endometrial circulation. Delivery of the egg became delivery of the endometrial membrane containing the egg. As gestation became completely intramembranous (intraendometrial) in primates, staunching and controlling endometrial hemorrhage at delivery or the end of the estrous cycle (menstruation) became prominent.

To be clearly understood the estrous cycle must be redefined. It is an abbreviated reproductive cycle. Physiology of the estrous cycle and physiology of the reproductive cycle are one and the same in mammals.

II

The Estrous Cycle and Reproduction in Infraprimate Mammals

The gradual lengthening of the estrous cycle of aplacental mammals into the lengthy gestation of placental mammals began with extension of the evolutional changes that were taking place within the egg. These were the evolutional changes already in progress and were simply a continuation of the changes by which mammals emerged from reptiles, i.e., greater consumption of oxygen for metabolism with a greater body heat byproduct and larger brain, the evolutional path upon which all terrestrial vertebrates have been advancing. In the mammalian lineage, however, this course of evolution seems to have advanced rapidly through development of a uterus in the mother and a placenta in the fetus. Although formation of the uterus and all associated function, structure, and behavior took place in extinct lineages, still the most important steps in the path can be roughly gleaned from studies of the roles of these new organs in reproduction of living mammalian orders.

In the past, the first terrestrial vertebrates reproduced in an aqueous environment through tiny eggs with limited yolk, similar to modern fish eggs. Like aqueous forms, they had small brains and small central nervous systems. The early embryos—after a larval or gill-respiring stage of development in water utilizing dissolved oxygen—completed their development on land, utilizing the greater source of oxygen in the atmosphere for their adult stages. Presently existing nonmammalian terrestrial forms are still committed in their early embryonic development to a limited oxygen source in egg yolk and albumin, reminiscent of the dissolved oxygen source of their ancient aqueous predecessors. However, they demonstrate further advancement of the evolutionary path. The seemingly retrograde evolutionary path of egg-yolk reduction in the mammalian lineage is actually a further advancing evolutionary step that removes the (yolk) barrier to greater oxygen availability in early embryogenesis. The egg in all vertebrates, aqueous as well as terrestrial, requires at least enough food yolk to carry the fertilized egg through its initial stages of cleavage and segmentation to the stage of its development at which it can absorb its own food. From the standpoint of early embryo formation, the internally incubated mammalian egg's liberation from the massive reptilian yolk body returned it to its originally aqueous environment with more available oxygen from maternal tissues than the dissolved oxygen in ponds or ocean. This additional oxygen available in the new internal environment enabled the egg to evolutionarily advance its early stages of embryogenesis, as well as its subsequent stages.

With increasing oxygen availability lasting throughout gestation, the evolutionary results to the embryo in early embryogenesis are a larger brain in the adult form. This is reflected in the progressive enlargement of the brain found in classes and orders of terrestrial vertebrates evolutionarily advancing in the atmosphere, in contrast to aqueous forms with brains and central nervous systems that have remained comparatively small despite evolutionary changes. Terrestrial vertebrates have been developing reproductive physiology that better utilizes the more readily available and unlimited oxygen source of the atmosphere. Other nourishment from maternal tissues, more abundant than stored yolk, has also enhanced larger brain formation in the embryo of mammals. The source of all body oxygen is dissolved oxygen collected from a wet membrane surface, whether the surface is skin, lung, or egg membrane. Wet membrane surface exposure to the atmosphere, however, collects more free oxygen than the dissolved oxygen it collects when submerged in the aqueous environment.

The trend toward larger brains in terrestrial vertebrates, accentuated and more rapid in recent forms of the mammalian lineage (primates), is a result of improved respiratory placentation. Nervous tissue functions in the temperature range somewhere between freezing and less than 107° F. It does not control its own temperature environment through any significant heat production in its own tissue, but by regulating the temperature of the body it innervates and controls. More body heat is generated in nonneural tissue than in central nervous system tissue despite the dependency of nervous tissue upon a constant oxygen source. Dissipation of excess body heat is governed by the central nervous system together with body morphology and physiology. It is the size, shape, contour, surface, surface appendages (scales, hair, sweat glands, feathers), and mechanical movement of the body that dissipate excess metabolically generated heat. Although some of these features generate metabolic heat, taken together with the morphological features, they are capable of dissipating more heat than is generated, depending upon environmental temperature. The great responsibility of the enlarging brain and central nervous system in mammals is to maintain its own environmental (body) temperature within the range in which it (nervous tissue) can survive and function. Nonneural tissues, in general, may sustain a higher, more prolonged temperature without damage, but sufficient protoplasmic damage and death of certain nonreplacing nerve cells result in unregulated body temperature that in turn can cause death of the animal.

The greater physiological delivery of oxygen to the brain and body of the mammal requires critical regulation of body heat loss. Body temperature regulation by heat loss is more efficiently accomplished at higher body temperatures than the surrounding environment, which then becomes a heat sink into which excess body temperature can be dumped more rapidly and efficiently. When environmental temperatures rise above or drop below the range at which the central nervous system can maintain its own immediate environmental body temperature, the central nervous system quiets the mammal's body functions that increase or lose heat too rapidly. This is the behavior and function in estivation or hibernation. During this period of body-function dormancy, the central nervous system increases body-heat loss or body-heat conservation necessary for its own preservation and survival. When this becomes impossible, the animal dies.

With constant and regulated high body temperature in mammals and birds, certain tissues other than those of the nervous system become vulnerable to the high body temperature. With body heat constantly gained and lost, not all parts of the mammalian body are maintained at the same temperature. Body morphology as well as the circulatory distribution of heat influence the temperature of different body parts. The heat generated by the muscular contraction of the centrally located heart is carried by the circulating blood peripherally to the skin and to the lungs, where it is cooled and returned. The testes hang outside the mammalian body for spermatogenesis at a lower temperature. Birds also have corresponding morphological arrangements to cool the testes during the breeding season. The greater adaptation of warmer bodied birds and mammals over their reptilian ancestors has resulted not only from the greater utilization of oxygen that produces the greater body heat, but the means of controlling the body temperature by regulating heat loss as well as its production. The body-temperature-controlling centers are reflex nervous centers that are located below the neopallium in the brain stem and are not large oxygen-consuming structures. Birds with small central nervous systems accordingly regulate their high body temperature through control of body morphology and function adapted for flight. Mammals, on the other hand, in physiologically acquiring more oxygen during their life history, have improved their adaptation through enormous enlargement of the neopallium with its larger reflex reserve.

The developing nervous system does not utilize a greater amount of oxygen at one stage and a lesser amount at a later stage of embryogenesis. As the central nervous system develops and enlarges, embryonic oxygen requirement and consumption irreversibly increase. Although the embryonic formation of a larger brain may advance according to the plan of structural elements in previous, less evolutionary advanced ancestral forms, the pattern for a larger central nervous system is there from the beginning. For example, a fully formed reptile or bird brain is not formed first, and then when more oxygen is made available later in gestation, additional nervous tissue is subsequently added to the brain already developed. Embryonic development of the central nervous system is progressive enlargement with progressively greater utilization of oxygen from the beginning of embryogenesis. Adequate circulatory establishment and oxygen delivery in the embryo must actually stay "just ahead" of needs of embryonic brain development in order to produce a larger brain. This is proved in bird reproduction.

During early embryogenesis in the bird egg, as in the reptile egg and early egg development (cleavage) in infraprimate mammals, the embryo derives its oxygen and other nutritional requirements from the yolk mass. In later stages of bird embryogenesis, its developing lungs remain dormant and the enlarging and expanding allantoic surface beneath the porous eggshell absorbs diffusing atmospheric oxygen and disseminates it through the embryonic blood circulation. The embryo, however, has already been committed in its early development to the limited oxygen of the yolk mass. Additional oxygen from the enlarging allantoic respiratory surface becomes available only in later stages of embryogenesis. The brain and central nervous system of the bird accordingly have remained small; birds have necessarily retained, and advanced only slightly, the small reptilian brain. The pallium, rudimentary in reptiles, is not significantly advanced in birds, if at all. Rather than evolutionarily advancing the brain and central nervous system, birds have greatly advanced oxygen consumption in the function and structure of their musculoskeletal system. This system forms and develops later in embryogenesis after the pattern and form of the embryonic brain and central nervous system are established. It was the huge yolk mass in reptile and bird eggs, initially required in terrestrial reproduction, that caused the large atmospheric source of oxygen to become available only in later stages of embryogenesis. The larger brain and central nervous system in mammals came about through earlier utilization of the oxygen source by eliminating the yolk mass in the egg. The allanto-chorionic-endometrial surface then developed earlier in embryogenesis and replaced yolk as the nutritional source of the embryo.

In embryogenesis the forebrain, the olfactory apparatus, forms first followed by the midbrain, thalamus, and neopallium. To develop the large mammalian brain the embryonic body had to be incubated, supplied its heat and heat regulation, and nourished to a large fetal size. A very large brain could not function in too small an embryonic body. Development and support of a large brain requires the mechanics of heat production, loss, and regulation of a larger developed body, a fetal body, that in turn requires a longer period of embryonic nourishment and incubation in its formation. This is evident in a comparison of ontogeny in aplacental mammals with ontogeny in placental mammals.

These advancing evolutionary stages in embryo formation and development continued and can be seen in the branching offshoots comprising the ascending orders of placental mammals. The evolutional advance is characterized mostly by steady enlargement of function and structure of the neopallium resulting from progressive improvement in circulation and respiratory nourishment in both early and late mammalian embryogenesis.

Elimination of unnecessary egg yolk and lengthening of the estrous cycle in development of more lengthy intrauterine egg incubation are initial manifestations of the mammalian evolutionary path that has continued rapidly in presently existing primates. In primates, the evolutionary path of these reproductive functions has become extended much farther, well beyond the maximum anatomical reduction in egg size, by means of prominently developed and critical endocrine physiology in the tiny egg. Gradual, progressive steps of this evolutionary path leading to primate reproductive physiology can be studied in the reproductive physiology of ascending infraprimate mammalian orders.

The beginning of an evolutionary path is evolutional change in progress so that no actual point or stage marks the beginning. A stage chosen as the first step in a path is arbitrary and for convenience only. In this instance the first stage or step chosen is the presence of a large yolk mass over which egg-membrane surface spreads and absorbs nourishment and oxygen to support embryo development. In the absence of a yolk mass in the advancing mammalian lineage there is complete cleavage and segmentation of the egg, with delay of embryogenesis; the embryo remains a knot of cells while the extraembryonic cells develop membrane surface to form placenta. Egg-membrane formation follows the reptilian pattern and order but, after loss of the large yolk mass, adapts in its spread to endometrial surface rather than yolk surface. Entoderm of the embryonic gut spreading over a large yolk surface absorbs nourishment, which its mesodermal covering, containing future mesenteric blood vessels, disseminates to other embryonic tissues. The same mesodermal circulation returns toxic, nitrogenous waste of embryonic metabolism for storage in the hindgut, a portion of which balloons into formation of the allantois. The surface of both these extraembryonic gut structures expands as the volume of their entodermal contents enlarge. Their surface contact with the chorion establishes embryonic respiratory surface.

In aplacental mammals, the egg remains a blastula suspended in uterine secretions. The outer egg-membrane surface, i.e., the exterior surface of the chorion, absorbs endometrial glandular secretions but does not attach to the endometrium as it does in placental mammals. The absorbed, or rather inbibed, glandular secretions mix with the yolk, enlarging yolk-sac contents and surface. In the egg of reptiles and early mammals, all embryonic metabolic waste was stored within the egg. The two enlarging entodermal storage sacs provide surfaces that accommodate the respiratory needs of the early embryo. The ovarian or early tubal egg in presently existing mammals delays embryogenesis until the blastula reaches uterine secretions. The egg at this stage of its development is about as small as it can be and still function. It is the largest cell of the mammalian body with only sufficient yolk to carry egg development to the blastula stage.

As the egg membranes gradually formed a placenta in their functions, external pouch or skin fold incubation of the embryo after hatching from the blastocyst became gradually replaced by lengthier uterine incubation of the egg before hatching, i.e., antepartum egg incubation in the uterus gradually encroached upon and replaced postpartum incubation of the previously externally hatched egg. The period of egg incubation was lengthened by the egg itself as it developed the endocrine functions necessary for delaying the maternal physiology that ends internal incubation, i.e., the uterine emptying functions that are the mechanical egg hatching (parturition) or estrous cycle–ending functions.

The evolutional changes in the mammalian egg, advancing its development through adaptation to the new internal body environment of the adult, occurred within the limits of mechanical hatching and delivery function and structure. Hatching and delivery mechanics became evolutional constraints when the entire incubation period of the egg became internal. In the egg that was laid in a nest or clutch the embryo pecked or cracked open the eggshell at hatching and emerged as a self-delivered neonate. When the entire incubatory period of the egg became internal, its delivery and hatching became part of basic tubal mechanics that move the egg. These delivery mechanics developing in the distal portion of the tubes, which were now becoming uteri, also became limits or evolutional constraints to advancing egg-nourishing functions. No egg, however well it adapted internally, or however well it was nourished, could evolutionally advance without hatching or being hatched into the exterior environment. Each adaptive improvement or advancement in egg incubation and nourishment produced corresponding adaptive changes or modification in hatching and delivery physiology, so that hatching and delivery functions in each order of mammals can be interpreted in terms of nourishing and incubation physiology.

The nourishing factors of the incubated egg that altered hatching and delivery physiology the most were the expanse of the egg surface, the depth of its endometrial invasion and attachment, and the proximity of the chorionic surface to the maternal blood circulation in the endometrium.

For example, the swelling, unattached blastula of Aplacentalia, freely suspended in endometrial glandular secretion that it imbibes into the embryonic yolk sac or gut primordium for embryonic absorption and metabolism, is haphazardly ruptured or hatched by pressure and friction of the birth canal during parturition. The eggs containing only embryos and fluid are squeezed from the uterus through the channel by uterine peristaltic action. The navel cords remain short, and peristaltic action of the tubular uterus hatches the egg by squeezing the embryo and its fluid through the blastocyst wall. The short cords snap all along the birth canal leaving the empty membranes strewn disorderly along the way as the embryos are delivered (Hartman, 1916).

In placental mammals, on the other hand, the blastula fixes and attaches to the endometrium for its nourishment. The navel cord must be longer than that of aplacental mammals, since the egg is completely opened and hatched gradually in steps before its delivery is completed. The fetus and its compartment in the egg of placental mammals are thus completely delivered into the atmosphere before the remainder of the egg is delivered. The egg membranes supporting fetal respiration must continue to remain fixed to the endometrium and function as the fetus hatches. Stages in mechanics of parturition are lengthier and more orderly in Placentalia, and are determined by internal egg (fetal) morphology as well as uterine mechanics that separate the egg membranes. A much larger fetus hatching through the chorion and the birth canal first requires more time, order, and specialized mechanics than those required to hatch a tiny embryo. A larger birth canal must be formed during which time the fetus must be sustained in its respiratory nourishment by a long cord attached to its fixed and functioning egg membranes. The cord "plays out" or "unloops" as the fetus is moved forward. The development of a large, oxygen-consuming central nervous system in the fetus of Placentalia accounts for the development of these parturition mechanics. As the fetus hatches into the atmosphere, the remainder of the fixed egg becomes functionless as it contracts in preparation for its separation and delivery.

Aplacentalia is the group of primitive mammals that once prevailed on the earth before placental mammals became dominant. They develop relatively small brains in large bodies when compared to adult placental mammals. Besides reptilian characteristics, such as possession of poison glands and more primitive body-temperature regulation, they hatch and deliver embryos after only days of incubation. Their oxygen source for all embryogenesis is from imbibed uterine secretions, since the blastocysts do not attach to the endometrium; consequently, their embryonic brain at parturition is very small. By hatching embryos after a short period of uterine incubation, the uterus and the maternal blood circulation are relieved of providing additional oxygen required during further incubation. Embryonic respiration is shifted to the embryonic lungs during further incubation, which is continued externally in the pouch. The embryos are well developed and mature in their foreparts at hatching and birth. Their small brains are matured sufficiently in that early stage of their development to enable the hatched embryo to seek out the pouch and move inside to a teat. The tiny lungs breathe, the brain and forelegs move the embryo through maternal fur covering to the pouch, and the mouth parts fix to the teat; its alimentary canal absorbs the milk nourishment. There is no return of these matured foreparts to their immature embryonic status for further formation of new structure and function. Once they carry the hatched embryo to find a nourishing teat in the pouch, embryonic respiration and nourishment during subsequent pouch incubation is through the matured and functioning adult structure within the embryo proper, rather than through the extra embryonal egg structures developed in placental mammals.

As these matured foreparts of the embryo in Aplacentalia continue to physiologically serve and support the embryo during pouch incubation, they grow without further embryonic development, reminiscent of growth in hatchlings of their reptilian ancestors. The hind parts, on the other hand, have no role in parturition and retain their undeveloped embryonic state throughout parturition and movement to the pouch. During pouch incubation the hind parts, as in reptiles of the past, develop to much larger size than the foreparts. Those reptilian foreparts that did enlarge grew in bone, muscle, and tooth tissue rather than brain and nervous tissue, which has further developed and enlarged during embryogenesis in placental mammals.

The lowest of the placental mammals, the pig, resembles in many ways the Aplacentalia. Its brain and foreparts, like the brain and foreparts of marsupials and momotremes, are not as advanced in their evolutionary stage of development as the brain and foreparts of higher Placentalia. The head is separated from the body by a short neck and, like Aplacentalia, the body and hind parts are relatively large. Reproductive functions of the pig closely resemble those of marsupials. The egg chorion as in Aplacentalia does not invade the endometrium. It attaches to the endometrium by simply lying against the endometrial surface, as the blastocysts in marsupials have been observed to do. While the blastula of the marsupial simply "sticks" to the endometrium as a small spherical bubble (Hartman, 1916), the blastocyst of the pig adheres as a cylindrical bag a meter in length. More eggs are often hatched and delivered of fetuses in the pig than there are teats, similar to egg production in marsupials.

Significant differences in reproduction between the pig and Aplacentalia are its lengthier incubation of the blastocyst and markedly delayed embryogenesis. Both of these differences result from gradual change of aplacental reproductive physiology to placental physiology. The more numerous eggs of marsupials initially suspended in uterine fluid become freely bunched like grapes, as they swell and develop within the uterus, which does not change in size. They adhere very slightly to the endometrium inside the uterine cavity as if they were lightly glued at one point on their surface. The eggs of the somewhat more advanced pig are carefully spaced at specific intervals, with their entire chorionic surface applied against the endometrial surface.

The uterus of Aplacentalia is a bilateral pair of organs that function independently of each other; the chambers of the Muellerian ducts do not fuse and connect into a single chamber, as they do in Placentalia. The uterus of the pig is a single-chambered organ of two horns formed by fusion of the lower ends of the ducts. In acquiring their spacing, pig zygotes "migrate" from one uterine horn to the other before attaching to the endometrium. The migration is most likely in fluid that passes through the connection between the two chambers. This period of time following fertilization, in which the pig zygotes become spaced at sufficient intervals for extensive egg-membrane expansion preceding placenta formation, is approximately the length of gestation and the estrous cycle in Aplacentalia. The developing pig eggs, initially in their free, unattached state are nourished by uterine fluid, which they imbibe through their yolk sacs similar to zygotes of Aplacentalia. However, there is a very marked delay in embryogenesis in the developing pig egg. In Aplacentalia embryogenesis begins even as the egg imbibes uterine fluid. This is a holdover from the reptilian characteristic of embryogenesis beginning while the egg is in the tube even before it is laid. In the thirteen days or so of uterine incubation of the aplacental egg, the embryo is formed and hatched; in the pig egg, embryogenesis, during this time, reaches only the primitive streak stage, i.e., embryogenesis has hardly begun. It is during this initial period of development that membranes of the pig egg enormously expand. This immediate and rapid expansion of egg membranes (chorion and yolk sac, i.e., entoderm and ectoderm) in the pig may be regarded as the modified blastocyst of Aplacentalia that grows larger and longer from yolk-sac nutrition before embryogenesis proceeds. This marked blastocyst enlargement and expansion in the pig, requiring adequate spacing in the tubular uterus, establishes a source of volume for the subsequently enlarging fetus, but it also serves other important functions, the most important of which is the establishment of an adequate fetal respiratory surface.

While surface for sufficient fetal gaseous exchange is provided with some respiratory waste gases passing back into maternal circulation for elimination, the flow of nutrition is mostly from the endometrium through the chorion into the circulation of the embryonic mesoderm. The entoderm and its covering of vascular mesoderm establish yolk-sac nutrition initially in many mammals both nonplacental as well as placental. It is the earliest most primitive form of embryonic nutrition and extends well into placental mammals before completely disappearing in primates. In all placental mammals any yolk-sac nutrition and yolk-sac placentation is replaced by more respiratory efficient allantoic placentation as gestation progresses. Like the case in reptiles and birds much of the metabolic waste of the pig embryo is kept and stored in the allantoic pouch rather than returned to maternal circulation, which is the case in upper Placentalia and primates. This waste-storage physiology diminishes with more efficient placentation, but continues up through Carnivora. Above Carnivora, in Insectivora and primate placentation, the allantois becomes only a vestigal pouch, a tube. The allantoic blood vessels dispose of all potential allantoic wastes by dumping them into the maternal circulation. After a short initial stage of yolk-sac nutrition, the enlarging allantois (from waste storage) in the pig embryo forms an increasing surface area inside the chorion, while the yolk sac forms a diminishing respiratory surface (from forming the embryonic gut) as embryogenesis advances. Mesodermal vascularization of the chorion necessarily shifts embryo respiration from the yolk sac to the allantois as embryogenesis lengths to fetogenesis. During embryogenesis, this entodermal yolk-sac respiratory surface shift to enlarging allantoic surface is seen in ontogeny of Reptilia, Aplacentalia, and infraprimate Placentalia. In upper primates the complete absence of the allantois shifts placentation entirely to the vascular allantoic mesoderm that renders implantation and placentation hemotrophic and respiratory from the beginning of gestation. This form of placentation in upper primates is the latest stage of the evolutionary path that began with decrease in size of the egg by evolutionary reduction in its yolk mass, continued as increase in egg surface, and finally became intimate trophoblastic invasion of the endometrial circulation.

In addition to providing a large volume and respiratory surface, the precocious development of the egg membranes in the pig also continues the pituitary support of the corpora lutea in the ovary. With a dozen or more fetuses simultaneously nourished in the pig uterus, at least that many corpora lutea function in the ovary. With gestation of the pig lasting only about sixteen weeks, estrogen and progesterone support of the pregnancy may be mostly, if not entirely, from the ovary, as in Aplacentalia. When fertilization fails to occur in the pig, degeneration of the corpora lutea occurs and estrus returns in about three weeks from the last estrus. Development and enlargement of fertilized pig eggs at about two weeks apparently begins to stimulate the pituitary to continue its luteotrophic maintenance of corpora lutea function. How fertilized eggs in the pig uterus are detected by maternal physiology is not clear. At this early stage the pig uterus is secreting its uterine milk and spacing the eggs, which rapidly become large enough for the uterus to space. Egg spacing appears to be a function of mechanical uterine peristaltic action. Ameboid movement in tiny enlarging blastocysts that can detect each other's presence and determine the intervening distance of one meter seems unlikely. The uterus, small at first, however, does grow more rapidly in length than in diameter and circumference. At the same time there does not appear to be secretion of endocrine substances by the imbibing egg into the maternal circulation (which is the case in the blastocysts of upper primates), as the chorion does not invade the endometrium. Egg products are absorbed by the pig endometrium when the egg dies in its earliest soft-tissue stages of development. In later stages, after fetal skeleton and hard tissues have formed, the fetus may be mummified and delivered along with the living fetuses at term and reported as "abortion at term."

Endometrial absorption of substances other than CO_2 from viable eggs of the pig is controversial. Some veterinarians (Roberts, 1971) theorize a luteolytic substance from the endometrium that blocks the action of the pituitary luteotrophic hormone on the corpora lutea, causing the latter to degenerate and end the estrous cycle. The presence of fertilized eggs in the uterus accordingly would suppress production of the luteolytic substance by the endometrium, permitting the luteotrophic hormone to maintain the corpora lutea and preserve the pregnancy. This may or may not prove to be the case in the pig. It would seem that rapidly enlarging pig blastocysts that can be detected and mechanically spaced within the uterus at two weeks should provide some sort of "tactilelike" stimulus in the uterus to which the pituitary responds by continuing its luteotrophic support of the corpus luteum. This may be the same type of hypothalamic-pituitary stimulation that occurs when suckling continues lactation.

If this is the case, continued mechanical enlargement and growth of the uterus from enlarging blastocysts would be the stimulus that maintains functioning corpora lutea and sustains pregnancy. Cessation of uterine growth from ripe placentas would end pituitary stimulation and bring on degeneration of the corpora lutea, with labor following. Levels of estrogens and progestins peak during gestation in the pig. The source of these endocrines is most likely the corpora lutea. The significance is that, like Aplacentalia, gestation in the pig seems to be entirely under the endocrine control of the corpora lutea. Whatever causes the corpora lutea to regress and cease their function ends the estrous cycle and early gestation in mammals. The corpus luteum regresses and degenerates when its endocrine support is withdrawn, whether it be pituitary support, anterior-pituitary-like support from the placenta (APL), or both, as in humans. There are many heterogenous forms of the same protein hormone that are difficult to correlate with change in physiological response (Hartree, 1989). In early human gestation placental hormones replace pituitary function in support and maintenance of the corpus luteum, but in later stages placental hormones also replace the endocrine functions of the corpus luteum. Although placental APL levels gradually drop, the corpus luteum is maintained into late pregnancy. Viable corpora lutea have been found in the human ovary at term. In the human, therefore, regression of placental endocrine function brings on labor, just as regression of the same endocrine functions from the corpus luteum ends the menstrual cycle often with cramps. In the pig, with pregnancy lasting less than half the length of human pregnancy, the more primitive condition of total corpus lutem control of gestation seems to exist. When pregnancy lasts longer, as in higher Placentalia, endocrine functions of the placenta replace the pituitary's endocrine support of the corpus luteum in order to lengthen pregnancy further. Eventually in upper primates and humans, the placenta even replaces the functions of the corpus luteum.

Compared to parturition in Aplacentalia, the labor and delivery function in the pig uterus is an orderly mechanical procedure that seems to be under neurogenic control. The most caudad fetus in each horn is delivered in sequence. Natural delivery of a cephalad-implanted fetus before the caudad-attached fetus is delivered would be mechanically catastrophic, if at all possible. The most caudal, longitudinal muscle fibers of the uterus shorten while the circular fibers distend and dilate. Each succeeding fetus in a horn passes through or over the implantation site of the previously delivered fetus. Each short-legged, short-necked pig fetus is bullet shaped and easily expelled mechanically. It is brought down to the perineum and vaginal canal by progressive shortening of the caudal, most longitudinal uterine muscle fibers that shorten the uterus and its horns beginning from caudal and spreading to cephalad regions. The average ten-inch cord length of the porcine fetus is thus adequate for fetal delivery before its placental membranes become detached and expelled. Pig fetuses, however, deliver randomly, in some ways resembling the haphazard parturition mechanics of Aplacentalia. During gestation, the lengthy egg membranes of the pig frequently fuse with each other, in which case fetuses could pass over the attached membranes of a previously delivered fetus. Several placental membranes may be delivered at intervals. Approximately one-quarter of the umbilical cords break during parturition in the pig; the rest may be delivered at intervals along with other fetuses and rupture after delivery (see Roberts, 1971). As in all litter-bearing mammals, there is no functional cervix in the pig uterus.

Involution in the sow uterus involves mostly the previously expanded endometrium, now redundant and collapsed. Like sterile pus, it is heavily invaded by leukocytes and resorbed. A retained placenta causes no ill effects and is apparently resorbed along with the redundant endometrium to which it remains attached. Since there has been no chorionic invasion or destruction of the redundant endometrium, there is no bleeding during reepithelization following leukocytic removal of the surplus membrane. Estrus can occur three days postpartum in the sow, but conception is rare if she is bred at this time. The next estrous period usually occurs after the young are weaned.

Most existing ungulates are survivors of slightly more advanced lineages than that of the pig. Typical representatives are domesticated ungulates such as the cow, horse, sheep, and goat; and wild forms such as deer and antelope, as well as more exotic ungulates such as rhinoceros, elephant, zebra, and giraffe. Not all mammalian forms have been subjected to detailed study. The evolutionary advancement or "higher" rank of these ungulates as evidenced by their increased adaptation (adaptation to wider and broader niches) has been more through specialization of body morphology, structure, and function than through neopallium enlargement and increased intelligence. The nervous motor control and coordination of specialized body structure were the basis of their improved adaptation. These hoofed mammalian forms developed longer legs with head usually separated from the body by a longer neck than that of the pig, without much greater enlargement of the brain. The adaptation of swifter flight and longer neck resulted in evolutionary modification of their reproductive function and structure compared to that of the pig. In particular, mechanical parturition functions were altered. The ability of the newborn fetus to stand on longer legs and run with the fast-moving herd very soon after its birth required a relatively larger, more developed fetal body, advanced in its development by a longer gestation period. The longer gestation was necessary for the development of neuromuscular coordination by the time of parturition. Incubational development accordingly became lengthened to acquire this feature or capability. Although a larger cerebrum and brain were not the most essential element in this form of evolutionary advancement, the growth and enlargement of the fetal brain during the latter portion of an extended incubation period did require more respiratory function in the maturing placenta.

Production of a larger fetal body more advanced in muscular coordination required much greater volume in the incubated egg, volume that in a litter would exceed even the limits of the mammalian uterus. Accordingly, an initial and significant evolutionary change involved reduction in the number of eggs produced at ovulation to a number that could be incubated and nourished within the uterus. Minor evolutionary change occurred in the uterus. This organ remained, in the pattern of its structure and morphology, essentially the same organ it is in the pig, an organ fitted for litter incubation, but the chambers of the fusing horns became connected, doubling the endometrial respiratory surface available to a single egg. Uterine function, especially its mechanical function in parturition, because of different fetal morphology, was modified. The production of one egg at ovulation for fertilization, the nourishment and incubation of one fetus to term, and the large but unsuitable double-horned uterine cavity shape were all results of evolutionary specialization of the mammalian body form in the fetus. These marked evolutionary changes developed first within the egg of ungulates before fully extending to the uterus, as they have gradually done in higher mammalian orders leading to primates.

To lengthen gestation and fetal development from the function of a single corpus luteum in a large-bodied mammal necessitated the alteration of endocrine physiology. Early stages of egg development in these more advanced ungulates are the same as early development in the pig, except the single blastocyst spreads more rapidly and extensively to cover the greater endometrial surface of both uterine horns. Early egg nutrition at this stage of development is from endometrial glandular secretion absorbed through the entire chorionic surface, amniochorion as well as the forming allantochorion. This imbibing form of blastocyst nutrition, intercotyledonary nutrition in these mammalian forms, ceases in the course of gestation when cotyledons and placentomes begin to form and function. After something like five or six weeks, depending upon the species, endometrial crypts and chorionic villi form the buttonlike placentomes, which are structures of closely associated allantochorionic vessels and endometrial vessels. When these structures begin to function they enter large amounts of gonadotrophin (APL in the mare) into the maternal circulation to preserve corpus luteal function. Later these structures also enter estrogens and progestins. In the cow they produce estrogens and progestins early on to maintain pregnancy. Follical stimulating hormone continues during pregnancy in the mare resulting in repeated ovulations and corpora lutea formation. The progestin from the repeated formation of corpora lutea in the ovary maintains pregnancy in the mare until the placentomes in the placenta furnish the necessary progestin level. In these ungulates the level of serum progesterone that results may prevent estrus from occurring during pregnancy.

Parturition involving the single large-bodied fetus at term in these advanced ungulates is more mechanically ideal when the entire fetus is located in one horn with only placental membranes extending into the other horn. When the fetal head lies in one horn and the body in the other horn, wry neck and transverse lie can occur. Other mechanical problems resulting from the unsuitable uterofetal morphology include hip lock and breech.

There is a definite cervix in the uterus, and cervical function is necessary in parturition mechanics of these single-fetus ungulates. Cervical function directs the mechanical force of uterine contraction to open the egg membranes by rupturing them while preserving their attachment and respiratory function, as the fetus is moved through the ruptured membranes to the exterior. In the cow and the pig the amnion fuses with the allantochorion; as in the case of humans, both membranes are opened by the same rupture. In the mare and some Carnivora the amnion containing the fetus is freely suspended within the fluid-filled allantochorion. The allantochorion is usually ruptured first through diffuse pressure controlled by the dilating cervix (external pressure on the egg), exposing some of the amniotic membrane surface. The delivering amnion, enclosing the fetus, later ruptures from pressure of the fetus from inside the egg similar to the parturition mechanics of Marsupial and the pig. The cord is long, permitting fetus and intact amnion to be mechanically moved, separate from the allantochorion. When the amnion covers the newborn fetus the latter may suffocate, since the ungulate dam may not lick and eat the amnion as quickly as the Carnivora dam may do.

Delivery of the placenta in these ungulates, unlike placental delivery in the pig, can be arduous and last several hours. The extensive size and surface of the huge placenta and the individual separation of each of the numerous placentomes make the third stage mechanically difficult and lengthy. When there are complications in domestic forms, delivery of the placenta make take two to three days and in some cases of retained placenta ten to twelve days (Roberts, 1971). Nursing the newborn can hasten delivery of the membranes by stepping up the rate of uterine involution, which separates placental cotyledons from the endometrial caruncles. Retained placenta or a portion of the placenta can and does occur but is not like retained membranes in the pig, without ill effects. Puerperal infections and passage of extensive purulent lochia can result. Normally, little lochia is passed. Something like one and a half liters accumulates within the uterus the first two days postpartum and is resorbed, in the same manner as the postpartum pig uterus functions. Similarly, normal lochia seems to be without bacterial contamination like the resorption in the postpartum uterus of the pig and the uterus of Aplacentalia. An infected or toxic lochia appears to be passed rather than resorbed.

Involution of the uterus and separation of the placenta are not separate events or functions in advanced ungulates. Involution seems, at least in part, to be neurologically controlled and regulated, as it occurs in peristaltic waves beginning at the tip of the horns, the exact opposite of involution in the pig uterus. In both pig and single-fetus ungulate, uterine peristalsis first expels the fetus and umbilical cord with a free unfused amnion, then continued peristalsis separates and expels the chorioallantois (and amnion when both are fused together). The collapsing placental membrane (chorioallantois) shrinks, contracts, and compresses into a volumetric mass as it separates. It must remain attached and retain sufficient fluid to keep it distended in providing functional respiratory surface during delivery of the amnion-enclosed fetus. In cases where separation begins before the amnion and fetus are delivered the fetus is born dead (suffocated). It is the unseparated caruncles and cotyledons not the intercotyledonary placental surface that must continue the respiratory function of the fetus during delivery. The two fluid compartments, the amnion enclosing the fetus positioned inside the allantochorion, may be separate compartments of fluid, or the amnion and chorioallantois may be fused into a single compartment as in the human. In either case, respiration of the fetus through the chorioallantois cotyledons or placentomes must be maintained during fetal delivery.

Around the tenth day postpartum, the lochia may become bloody from involution of the endometrial caruncles. In the cow, the caruncles are endometrial excresences of epithelially denuded endometrial vascular tufts that are convex projections with their stroma invaded by concave-shaped tufts of chorionic cotyledon vessels, the projection of which closely associates the vessels of the two circulations. In the ewe, the geometric configuration is opposite. The endometrial caruncles project concavely like cups into which convex chorionic vessels radiate. After the first postpartum week or so, when resorption of lochia and involution reduce the size of the uterus, apparently the stalk of each caruncle is constricted, and the caruncles become autolyzed and bleed during their healing, which is complete in about a month. The mare can be bred as early as nine days postpartum, indicating rapid uterine involution. It has been suggested that a silent estrus may occur in the cow; a productive estrus, however, usually occurs after the calf is weaned.

The placenta of Carnivora is not much different in microscopic structure from that of ungulates (see Grosser, 1909). Its different function, however, results from, and can be seen in, evolutional changes in the earlier stages of structure and function in the developing Carnivora egg. Whereas the spaced blastocyst in the ungulate rapidly elongates to cover the available endometrial surface before it fixes, the enlarging blastocyst in Carnivora fixes sooner, while it is still small and retains its globoid morphology. The spherical chorion within the tubular horn contacts and destroys the endometrial surface in an equatorial band or zone as it attaches.

In contrast, the egg chorion of single-fetus ungulates, after its first duty of acquiring surface and volume, destroys the surface epithelium of the endometrium later in pregnancy to fulfill its second duty of prolonging the incubation and gestation period. This destruction of the endometrial surface epithelium to provide the "injection" of chorionic endocrine sustenance necessary in prolonging pregnancy also opened up a vastly superior system of nutrient and metabolic exchange between maternal and fetal circulations. The huge chorio-endometrial surface contact resulting from provision of greater chorionic volume became not only unnecessary but a detriment in adequate fetal-maternal exchange in the placentomes. Accordingly, during pregnancy in single-fetus ungulates, the intercotyledonary surface discreetly ceases its exchange functions, relegating all nutrient and respiratory exchange to the discrete areas that form cotyledons. In doing so, enough but not too much, oxygen and respiratory exchange is afforded the fetus. Where endometrial and chorionic surface is diseased, as in case of infection, more cotyledons (adventitious placentae or accessory placentomes) necessarily form over undiseased surface areas. In some cases the undiseased placental area of the chorion may be so limited and closely filled with functioning cotyledons that the cotyledonous area of that particular ungulate placenta takes on the appearance and texture similar to the cotyledonous discoid placenta of primates. The number and size of the placentomes in normal single-fetus ungulates are a measure of the effective respiratory area and exchange. There are approximately 70 to 120 placentomes in the cow placenta and 80 to 90 in the ewe (Roberts, 1971).

This endometrial destroying and invading function of the chorion provides a histotrophic form of egg nutrition. It is slight and occurs late in pregnancy long after egg fixation in single-fetus ungulates. It occurs earlier during attachment and is deeper and more lengthy in Carnivora. During the earlier attachment of the Carnivora egg, histotrophic nutrition replaces more of the imbibition of endometrial glandular secretion and yolk-sac nutrition of the chorion. It can be seen that the histotrophic stage of nutrition, late and less extensive in ungulates, becomes earlier and more extensive in Carnivora. The evolutionary changes inside the egg are structural and functional changes that shift egg nutrition from oxygen-limited yolk, endometrial glandular secretion and histotroph, to more adequate respiratory exchange in the endometrial vascular bed, from entoderm to ectoderm, from vascular mesoderm covering yolk-sac entoderm to the vascular mesoderm covering the allantois and lining the chorion. This more adequate respiratory exchange resulting earlier in egg development fostered the earlier enlargement and function of the embryonic brain and central nervous system in the Carnivora embryo. The allantois, the embryonic structure about which the switch in nutrition pivots, in Carnivora retains some of its original function as a storage organ of embryonic waste, just as it does in orders below Carnivora. Its enlargement during embryogenesis is the basis for the embryonic shift of respiratory circulation from blood vessels in the mesoderm covering the entoderm of the diminishing yolk sac to vessels in the expanding mesodermal covering of an enlarging allantoic chamber. The chorioallantois in Carnivora not only destroys the endometrial surface epithelium early, but it also invades and penetrates the endometrial stroma in reaching to but not into the endothelium of endometrial vessels.

This deeper attachment of the Carnivora egg, in producing the zonary or band placenta, brings about alteration in parturition, but the alteration is not marked and striking. The mechanical functions are more orderly than the irregular deliveries of fetuses and placentas in the pig. Since the deeply invaded band-shaped placenta is more adherent to the endometrium in Carnivora, the lowest placenta in each horn must be separated and delivered after emptying its fetus before the next adjacent fetus and placenta can be delivered from that particular uterine horn. After the placenta is detached and passed, the denuded and damaged area of attachment must dilate and allow the next fetus and placenta to pass through. This is mechanically feasible because the invading chorionic villi invade stroma down to but do not enter endometrial blood vessels. The villi only lie against the capillary endothelium. No hemorrhage occurs when the placenta separates.

The lengthy horns of the uterus are retained in Carnivora, multiple eggs are incubated and nourished, and involution or labor begins caudally, as in the pig. Like the uterus of the pig, there is no cervix, but the poles of the attached eggs at term are free of villi. The ends of the chorions do not fuse with adjacently attached eggs as they may do in the pig. The amnion, freely suspended within the chorion, remains unfused with either chorion or allantois except in its reflection at the umbilical stalk. As a result, the Carnivora fetus and cord are delivered enclosed in the amnion, which is attached to allantochorion (the placenta in formation) by the umbilical cord. The lochia may be amber or green in the early postpartum period, but toward the end of the postpartum period (one month) the lochia is clear mucous. After six weeks or so a brown-pigmented ring may mark the site of each implanted egg. Estrus occurs usually after the young are weaned.

This improvement in earlier embryonic respiratory exchange afforded by the deeper embedded egg over a smaller endometrial surface area accounts for further evolutionary advancement and enlargement of the embryonic neopallium in Carnivora, Chiroptera, Insectivora, and other mammalian orders above ungulates. The time or stage in development that the egg reaches before it fixes to the endometrium, i.e., the extent of respiratory egg surface developed at the time of fixation, determines the type of placenta. The type of placenta, in turn, determines the final parturition or estrous cycle–ending function of the uterus. The blastocyst starts its growth and enlargement through absorbed secretions before it attaches. With first stages of egg growth and enlargement taking place when the egg is free in the uterine cavity, the endometrial development at the time of attachment is similar in all the infraprimate orders. The squeeze of the tubular uterus in lieu of attached eggs in the estrous cycle constricts and renders ischemic the endometrial surface, resulting in mucosal damage. In primates, however, the egg fixes still earlier in the cycle than it does in Carnivora and its chorion invades deeper. The primate egg differs from the infraprimate egg in that it only starts growth and enlarges after attachment, and the removal of the deeper, more extensively damaged endometrial surface markedly alters the primate estrous cycle. The hemotrophic nutrition early developed in the upper primate egg to support still earlier development of a still larger embryonic central nervous system structure is the basis for metaestrous bleeding during the leukocytic removal of the residual endometrial slough that ends the upper primate cycle.

Length of the initial period of glandular nourishment in all infraprimate mammals diminishes in progressively ascending orders. Although the egg begins to fix or attach to the endometrium at progressively earlier stages, it does not damage or enter endometrial blood vessels at any time during gestation. Mechanical separation and delivery of the egg membranes and removal of damaged endometrium in the reproductive cycle do not involve bleeding from broken or damaged endometrial blood vessels. The egg attaches only to endometrial surface or invades only endometrial stroma. When fertilization fails to occur, the abbreviated endometrial cycle is ended in its early stages by the mechanical squeeze of uterine delivery, which becomes applied to the egg-fixing surface of the endometrium. The very early developed membrane surface and stroma thus damaged by parturition mechanics is too small in volume and mass and too undeveloped to be mechanically delivered. Damaged portions of the endometrium still residual and attached are removed in the postpartum period by leukocytes that invade the endometrium and absorb the debris. Because the ovarian, tubal, and initial uterine egg in all these infraprimate mammals is yolk- and gland-secretion nourished, removal of the discarded glandular endometrium in the metaphase of their estrous cycles is essentially removal of damaged stroma with intact endometrial blood vessels. The estrous cycles in these orders are thus remarkably similar, their terminations differing from those of primates in being characterized by prominent leukocytic infiltration of the damaged endometrium without bleeding.

Summary

Early stages of the estrous cycle established in extinct terrestrial vertebrates, transitional between reptiles and mammals, involved miniaturization of maternal reproductive structure and function that accompanied reduction in ovarian egg size. These evolutional changes were prerequisites for the later evolutional changes that have taken place at the end of the lengthened cycle in infraprimate mammals. The evolutional changes at the end of the cycle involve marked changes in uterine mechanics as the egg developed its placenta.

The egg in its nutritional adaptation first adjusted to the tubular chambered organ developed to move sperm internally and the egg externally. It increased its nutritional surface to the limits of the newly forming uterus by reducing its numbers to a single incubated egg with communication of the double uterine chambers to form a single nutritional surface. At the site of its fixation to the endometrial surface, it began to invade the endometrial surface and attach for endocrine extension of the incubation period. The chorionic invasion of the endometrial stroma opened up greater respiratory exchange. As endometrial invasion began earlier in gestation, it also became deeper in a more limited surface of the endometrium. Separation of the attached egg invoked special squeeze mechanics in the incubatory uterus. These special squeeze mechanics account for the endometrial damage in the estrous cycle. Finally, complete merging of the primitive double chambers of the infraprimate uterus into a single-chambered, smooth-walled organ occurred along with more specialized endometrial squeeze mechanics to form the primate uterus.

Since the egg at the beginning of the reproductive cycle in infraprimate and lower primate mammals utilizes yolk and glandular secretions before it fixes and attaches to the endometrial surface, the endometrial changes in the abbreviated cycle from failed fertility, the estrous cycle, therefore, is similar in all these mammalian forms. The uterine squeeze that concludes gestation, i.e., the squeeze that separates the attached egg membranes or placenta, in the estrous cycle squeezes and damages only the superficial endometrial stroma that was destined to furnish egg nutrition and enter into placenta formation. Because the damaged stroma is so early in its development in the estrous cycle and because it is more cellular than vascular in comparison with the endometrial stroma of upper primates, it does not separate or cleave and cannot be mechanically delivered by the uterus. The damaged stroma is resorbed by infiltrating leukocytes from the injured endometrial capillaries. The damaged endometrial capillaries of the less vascular surface heal without bleeding.

When the embedding egg invades the endometrial blood circulation in upper primates endometrial hemorrhage becomes an added result of endometrial damage in the estrous cycle.

III
Development of Menstruation within the Estrous Cycle of Primates

The evolutional path of fetal brain enlargement that has produced the various orders of infraprimate mammals has continued on to produce primates. Existing primates represent small, short, closely related steps in the evolutionary path that are separated by narrow gaps. The narrow gaps are created by extinction of some forms through natural selection, allowing more convenient classification of the various surviving groups according to their kinship and origin. Evolution is recognized as an irregular course of survival in varying and branching forms that, in general, show more differences from their main ancestral stem the farther they branch from it. Extinction in variations of a form creates its branches. The longer a branch survives the fewer relatives and offshoots it is likely to have as extinction reduces their number. More recent forms, i.e., branches closer to the main stem, like primates, tend to have more numerous and closely related kin. Evolutional steps are not in direct, straight lines but form a zigzag course in all directions about a main stem. Although straight in its direction, the main stem, in this case progressively greater oxygenation of the mammalian fetus, may contain few existing species, since most surviving forms are only branches or offshoots.

Existing primate species are a cluster of two hundred or so relatively recent forms, an order of mammals showing numerous variations in their most recent forms despite relatively minute differences in their reproductive physiology. Evolutional steps to the human are obvious from studies of the brain and central nervous system in primate adults and fetuses. However, small increments of improvement in egg, embryo, and fetal respiration are less obvious in the placentation physiology that produces them. The evolutionary path in primates can be discerned in structural change that continues to increase functioning respiratory surface of the egg membranes earlier and throughout gestation. The smooth chorion lying against the endometrial surface progressively destroys deeper endometrial layers until it reaches endometrial blood vessels of the maternal circulation. It then forms blood channels or lacunae within cords of chorionic cells, which advance to villi projecting into extensive blood sinuses. These are the evolutionary steps that pushed forward respiratory nourishment of the mammalian egg to hemotrophic nutrition in upper primates. Hemotrophic nutrition in placentation of upper primates induced marked change in uterine egg hatching and delivery functions, mechanical functions that developed at the end of gestation. The combination of these nourishing and mechanical functions in development of the upper primate uterus accounts for menstruation in the primate estrous cycle.

Because the steps of this path are so close and so similar in the numerous primate forms, the evolutionary course can be discerned and studied more easily in representative forms from wider spaced primate groupings that reveal the advancing steps more clearly. The primate groupings considered are prosimians, platyrrhine monkeys (New World monkeys), catarrhine monkeys (more recent Old World monkeys), anthropoid apes, and humans.

Through many forms of terrestrial vertebrates, the vascular and respiratory egg mesoderm, covering entodermal epithelium of the yolk sac and allantois in large eggs, has become the dominant and precociously developed primary germ layer in the tiny egg of upper primates. In upper primates the mammalian egg enlarges its respiratory surface area of endometrial contact through precocious growth and expansion of its mesoderm and ectoderm before initiating embryogenesis. Embryogenesis is initiated only after these membranes acquire adequate surface respiratory nutrition for subsequent central nervous system formation and growth, whether the respiratory surface is a smooth or trabecular membrane or a villus bathed by a flowing bloodstream. Adaptation of the internally incubated egg to its environment induced corresponding change in mesodermal structures of the maternal adult. Modifications of tubal epithelium and muscularis in forming uterine function and structure reached the limits of the maternal environment in primates; the maximum available oxygen in mammalian body tissues is utilized by the egg in the human and upper primate reproduction.

From the evolutionary course of mammals up to the present, one might predict the next evolutionary step beyond present human physiology: more precocious development and enlargement of the embryonic heart and mesodermal egg membranes for even greater embryonic respiration, and more delay in embryogenesis and embryonic central nervous system development until still more extensive respiratory surface and function is established within the endometrium. While these changes would increase early embryonic respiration, corresponding changes might also become necessary in the adult cardiorespiratory system and circulation to support fetal respiration in the latest stages of gestation. A larger lung surface and a larger heart may become necessary as well as a body that utilizes even more oxygen. The female of such future human forms would menstruate in her reproductive years not much differently from menstruation in existing primates. There does not appear to be a trend toward returning to a histotrophic nutritional stage in early primate ontogeny. The advancing development of the mammalian nervous system, presently continuing in upper primates, extended into the future, mandates a continued increase in oxygenation of the developing embryo throughout gestation, including both early and late stages. Fertilization of a small egg and its subsequent attachment to an erythemic endometrium do not appear likely to change.

As more embryonic brain tissue is formed during embryogenesis, it permanently obligates a larger increment of the available oxygen supply. The largest embryonic brains develop in primates and require the largest oxygen supply. While all other embryonic tissues also require oxygen, there can be energy storage in nonneural tissues in the form of fat, glandular secretion, yolk and albumin, and other cellular products. Muscle can operate temporarily in a state of "oxygen debt." Brain tissue, on the other hand, functions only from a constant supply of free oxygen, i.e., chemically unbound oxygen. Other body tissues can develop under conditions of limited and varying oxygen supply and sustain a brief interruption of oxygen supply. When a temporary blockage of blood circulation and oxygen delivery to the fetus occurs, such as that produced by brief pressure on the umbilical cord, the brain and central nervous system may be the only tissues to suffer irreparable damage. The embryonic and fetal brain and central nervous system may be regarded as holding the "right of first refusal" on the body's "free" oxygen supply in a circulating bloodstream. A study of structure and physiology of fetal blood circulation indicates specific divisions in the pathway of blood circulation and oxygen supply to the embryonic brain and body. From the comparative and evolutional standpoint, however, the total oxygen supply covers the larger and immediate demand of nervous tissue first, with nonneural tissues developing and utilizing the residual. Fetal blood is neither completely arterial nor venous and has a higher hemoglobin content than adult blood, which absorbs the atmospheric oxygen; the fetal blood maintains oxygen tension in a safe range that is sufficient to accommodate the needs of both nervous and nonnervous fetal tissues.

During development of the egg, the embryonic oxygen supply in all terrestrial vertebrates is an obligation of the specialized egg membranes, which are extensions of embryonic germ layers. In the developing embryo, blood, heart, and great vessels of the head form early along with the brain and distribute available oxygen to embryonic tissues. The only oxygen that becomes available, however, is the increasing amount of oxygen absorbed by the structural and functional expansion of the egg membranes. In mammals, with the exception of monotremes, the egg adapts to the uterine environment in all its stages of development during incubation, but can only adapt to the limits to which oxygen availability can be progressively increased during gestation (see previous chapter). The fetus does not store free oxygen for later use but only satisfies its immediate oxygen needs and the needs of its developing central nervous system.

Fusion of the uteri, seen as bilateral, separate organs in aplacental mammals, began in those extinct mammalian forms transitional between aplacental and the lowest placental mammals. Movement toward fusion of the Muellerian ducts started the combination of separate uterine cavities into a single uterine chamber, adding greater chorio-endometrial surface contact for respiration of the single large uterine egg during its development. First placentation was the simple epithelio-chorial type exemplified in the pig in which expansive epithelial surface contact was necessary in furthering fetal respiration. With marked improvement in respiratory exchange accompanying specific cotyledonous areas of superficial chorionic invasion into the endometrial surface (exemplified in advanced single-fetus ungulates), the extensive, anatomically smooth surface area of chorio-endometrial contact (exemplified in the pig) became unnecessary. The respiratory surface area of the chorion became transferred to the more efficient focal, chorionic areas (dispersed cotyledons).

Further fusion of the two separate uterine cavities also was no longer necessary to further respiration; consequently, fusion of the Muellerian ducts remained incomplete in evolution of placental mammals up through Carnivora. The extent of these specialized cotyledonous areas of the chorion was not limited by the function of their hormone production but covered the chorionic surface only to the extent increasing fetal respiratory demand became satisfied, not too much oxygen and not too little. The evolutional constraint was respiratory, not endocrine. Further oxygen demand of the mammalian fetus in late stages of development in orders above Carnivora brought on completion of the fusion to form the single-chambered primate uterus. The fusion concentrated the blood supply and oxygen delivery to the single, smaller uterine chamber.

Embryo development and embryonic brain enlargement during gestation, utilizing only that increasing amount of oxygen required in formation and development of brain and body, are genetically determined. The environment of the internally incubated primate egg, created by fusion of the Muellerian ducts, is also genetically determined and controlled. The development of the cardio-respiratory system of the adult that delivers atmospheric oxygen to body tissues is similarly determined by genetic formulae. When the ancestral egg began adapting to the internal body environment, its early expressed genes became constrained by later expressed genes that produced the oviducts and uterus. Both genetic expressions, those of the early developing egg consuming more oxygen and those that produced the Muellerian tissues that delivered it, were limited by the final genetic expressions, producing the adult cardio-respiratory system that collects and distributes atmospheric oxygen. As evolution is gradual change in the hereditary formula of the species, the egg, or embryo, adapted in minute, gradual steps over evolutionary time to make greater use of a resource available earlier in its development. Internal incubation provided the egg an environment genetically determined and controlled, an environment the egg itself controlled, an environment that is the egg itself in later stages of its development. Whereas the externally hatched egg adapted to a fickle environment it had no control over, the internal egg adapted to a steady, even environment that actually responded to its adaptation. The egg adapted to the endometrial surface, and later in its adult stage its own endometrial surface adapted to the implanting egg. The results were genetic modification of maternal tissues and organs reached at the end of gestation and genetic adjustment of the fetal brain/body ratio to a higher value at the end of gestation.

Although uterine functions are intimately associated with incubation and nourishment of the primate egg, the uterus is not essential for either, as proven by human ectopic pregnancy. The human egg can implant, nourish itself, and be incubated within the abdomen while located outside the uterus, and, in certain cases, even in the absence of the uterus. Such an egg in its extrauterine location, of course, cannot be naturally hatched to the exterior; delivering and hatching the egg is the primary role of the uterus in all placental mammals. In its normally implanted location within the uterus (just as it does when implanted outside the uterus), the primate egg provides its own nourishment through its membranes and the placenta they form. It also hatches itself, through the uterine function that it controls when it is located normally within the uterus, just as the egg of other placental mammals does (the ovary, rather than the egg, controls uterine function in aplacentals; see chap. 2). During the reproductive cycle of placental mammals, the egg acquires total control over the uterus and its emptying functions, indirectly through ovarian hormones as the egg matures in the ovary, as well as directly through the mechanical influence of the egg's physically enlarging presence within the organ. The tiny primate egg, in controlling uterine function through its indirect endocrine influence, begins with replacement of pituitary function that controls the ovary. This total control of uterine growth, function, and structure by the egg during the reproductive cycle is the heart of primate evolution, i.e., the organ's structure developed from the mechanical and functional needs of the egg during parturition, not from its nutritional needs during incubation.

The ovum of the pig, the lowest of the existing placental mammals, may be looked upon as an egg that sits atop the endometrium with all smooth, or avillous, chorionic surface, free and in contact with the uterine lumen similar to the egg in aplacentals. Its fixation is simply an interdigitation of the chorionic membrane with the endometrium. No invasive attachment occurs throughout gestation; the cleavage plane of separation at parturition is the potential uterine lumen or cavity space between chorion and endometrial surface. Egg nutrition from endometrium to chorion must pass through this potential uterine cavity space with most of the metabolic waste of the egg being stored in the egg rather than passing backward through this space for absorption by the endometrium. In single-fetus ungulates this form of placentation occurs initially, but the ovum erodes the surface epithelium of the endometrium in patchy areas of attachment in late stages of gestation. In the patchy areas of attachment the uterine cavity is obliterated with erosion and loss of endometrial surface epithelium in formation of endometrial caruncles. The Carnivora ovum completes this initial, smooth, free, chorionic growth period quicker and invades deeper in the endometrial stroma, but not into the endometrial blood vessels. The ovum of lower primates passes through an even briefer smooth, chorionic growth stage, begins endometrial attachment still sooner, invades stroma even deeper; in upper primates the egg immediately opens and enters the finer endometrial, blood, vascular channels (Wislocki, 1929 and 1930).

The egg of Carnivora, like the egg in orders below it, enlarges spherically from uterine secretions. It fixes and attaches to the tubular endometrium after it reaches a considerable size (see chap. 2). It begins its histotrophic nutrition much sooner than the egg in ungulates, but much later than in primates. It represents the intermediate evolutionary stage, between late-ungulate egg fixation and early-primate egg attachment. In the tubular horn of Carnivora, the enlarging spherical egg makes endometrial surface contact over a ring or band-shaped area of the chorion, in which the trophoblast invades down to and against the endothelium of endometrial blood vessels. The egg poles projecting into the tubular lumen on each end are the bare, smooth areas of chorion devoid of invading trophoblast or cotyledons. In the fused horns forming the uterine chamber in lower primates, the egg also contacts the endometrium, leaving the uncontacted egg surface smooth and bare. In this chamber the spherical egg contacts and invades the endometrium over its lower pole initially, leaving the upper chorionic surface that projects into the uterine lumen bare and free of invading trophoblastic villi. The endometrium on both anterior and posterior uterine walls proliferates and forms a thick, endometrial pad, designated a "placentoid," which lies both beneath and opposite the site of egg attachment. The surface of the bulging, chorionic egg sphere may contact the opposite placentome, invade it with trophoblastic growth, and establish a bilobed placenta. This bilobed placental form in lower primates is a step in evolutionary advancement of the placental form from the Carnivora band type in the tubular uterine chamber to the discoid placenta in the spherical chamber. It represents an adaptation of egg development to the mechanically and geometrically different shaped environment, which the primate uterine chamber affords. In some primates the egg may not invade the endometrium of the opposite placentome, forming only a single-lobed placenta at the primary attachment site. When this occurs the opposite, unutilized placentome disappears during gestation; the proliferated glandular and stromal endometrium regresses (or is resorbed), and the glands, as in other parts of the endometrium, cease their secretory function and become inactive as gestation progresses. A single individual primate female can develop a single-lobed placenta in one gestation and a bilobed one in a subsequent gestation.

The histotrophic stage of mammalian egg nutrition that starts late in the gestation of single-fetus ungulates with erosion of endometrial surface epithelium starts earlier and extends deeper into and through endometrial stroma in carnivoral gestation. It begins still earlier in lower primates, while the egg is still particulate in size. Sheets and meshes of chorionic trophoblast and mesodermal cells of the lower primate egg erode the endometrial surfaces and engulf stroma within the placentoids. The pabulum, or "embryotrophe," of glandular secretion, digested cells, intercellular or edema fluid, and blood cells is formed. This is the nutrition that the egg of lower primates utilizes as it embeds and attaches to the thickened, swollen, and edematous endometrial placentoids. This is histotrophic nutrition, sacrificed endometrial cells and stromal cell bodies added to the glandular secretion and procured as food by the growing and developing primate egg in its initial attachment to the endometrium. It is food that only replaces glandular secretion that the mammalian egg has evolutionally lost. It is also food that contains sufficient oxygen for egg respiration during early embryogenesis in lower primates. The initial metabolic requirement of the particulate egg in these mammalian forms is similar in that they all require yolklike pabulum for beginning embryogenesis. The nutritional and metabolic need of the egg upon its first contact with the endometrium in these forms is destroyed endometrial structure. This special structural characteristic of the endometrium, established at the beginning of the mammalian reproductive cycle, results in the terminating phagocytic infiltration and phagocytosis that distinguishes the end of the estrous cycle (metaestrus) in these primates as well as infraprimate forms. The mechanical parturition squeeze of the uterus, that delivers and separates the egg from its endometrial attachment at the end of the estrous or reproductive cycle in infraprimate mammals, squeezes and damages only the superficial cellular layers of the endometrial stroma in upper primates and the human. This is the surface egg-contacting tissue of the endometrium that is damaged, destroyed, and utilized by the attaching egg. As the superficial endometrial layer is a cellular, pabulum-providing tissue in lower primates and infraprimates, its response to the squeeze injury is only phagocytosis absorption of damaged cells, not bleeding.

The microscopic studies of the placental attachment in lower primates show a functionally deeper attachment in placentoid cushions of the endometrium relative to Carnivora placentae but a more shallow attachment when compared to the vascular attachment of the placenta in catarrhine monkeys, the anthropoid apes, and humans—the primates that menstruate. Beneath the placenta of lower primates, the endometrial layer next to the muscularis contains the fundic portion of endometrial glands that at the placental margin extend into full-length glands with openings bridged by smooth chorion. There are no anchoring villi, and the trophoblast is entirely syncytial; cytotrophoblast is absent. The advancing sheets of trophoblast, spreading later in gestation than in upper primates before piercing endometrial blood vessels, form lengthy, labyrinthine blood channels. The lining trophoblastic trabeculae contain mesodermal cores; villi do not form. No implantation bleeding corresponding to the "placental sign" in upper primates occurs in these primates. They are considered to rank, in their placentation, between Carnivora and Insectivora on the lower, more primitive side and lower catarrhine monkeys on the more advanced upper side (Wislocki, 1930).

Hematrophic circulation established by the egg in superficial layers of the endometrium following a stage of histotrophic nutrition occurs in platyrrhines. Structure of the endometrial membrane in the placentoids at the beginning of the reproductive cycle in this group of primates is the same type, with a character similar to the endometrial membrane of infraprimates. Hemotrophic nutrition established sooner in deeper endometrial layers occurs in the more advanced catarrhines. In catarrhines as in humans, the endometrium does not proliferate into placentoid cushions. In this infraorder, Catarrhini, that includes the human, all surface of the endometrium thickens equally. In humans and anthropoid apes, all nutrition throughout gestation may be regarded as hematrophic and respiratory rather than histotrophic. Arterial capillaries proliferate more superficially in the compacta beneath the epithelium, distinguishing the surface membrane from that of lower primates and infraprimate mammals. As the egg's early formed trophoblastic villi invade stroma, they reach the endometrial capillary layer sooner and pierce the endometrial blood channels. Anchoring villi penetrate and attach to the base layer of the endometrium and a layer of fibrin separating the endometrium basalis and the placenta is laid down. The endometrial capsularis above and the endometrium basalis below confine expansion of egg villi within the superficial deciduate layer of endometrial capillaries. This endometrial layer in menstruating primates is the delicate subepithelial capillary network structured to establish a microscopic placental circulation around an embedding egg particle. As maternal blood enters the lacunae of the tiny trophoblastic syncytium, it is prevented from clotting in order for the lacunae to grow into more extensive circulation channels. In nonmenstruating primates this endometrial layer is a proliferated edematous cellular cushion. These differences in the structural characteristics of the surface endometrium when the egg implants determine whether endometrial phagocytosis or endometrial phagocytosis with bleeding will mark the end of the estrous cycle.

Increased delivery of oxygen to the embryo early in primate embryogenesis is reflected in progressively earlier embedding of the egg membranes, i.e., trophoblastic invasion of the endometrium mentioned above. This increases the chorio-endometrial surface contact functionally more than structurally. The histotrophic stage of egg nutrition, which begins late in ungulate gestation, begins earlier and lasts longer in Carnivora, absorbing more tissue oxygen all the while from the growing and expanding endometrial tissue. The blood supply to the separate uterine horns of the Carnivora uterus is adequate for egg and fetal respiration and nutrition, and the organ's delivery and hatching mechanics are adequate for the depth that carnivoral eggs fix or attach to the endometrium. The greater depth of trophoblastic invasion taking place even earlier in primates, however, has resulted in marked evolutionary changes in both hatching mechanics and egg nutrition. Without (trophoblastic) damage or invasion of endometrial blood vessels, the eggs of Carnivora can be mechanically separated in the unfused uterine horns at parturition in the same manner ungulate litters are born. Each egg can be expelled over the attachment site of the previously delivered (lower implanted) egg, without hemorrhage and without cervical function. Greater egg respiration from earlier and deeper egg attachment in orders above Carnivora has necessarily effected very marked evolutional change of parturition mechanics. Deeper egg invasion brought about evolutional change in uterine structure and function, which accomplishes safe mechanical delivery of the deeper and dangerously embedded egg. Among the changes: The number of eggs ovulated were again reduced to one, as in the case of single-fetus ungulates; cervical function developed; greater more complete fusion of the Muellerian ducts produced a smoother and more regular, uncompartmented uterine chamber with a much greater blood circulation from the resulting blood supply; and the endometrial surface converted from a glandular, histotroph-producing membrane to a more vascular respiratory one. It was the necessary change in parturition mechanics of a deeper embedded egg that brought on more complete fusion of the Muellerian ducts to form the primate uterus.

These evolutionary priorities have persisted not only up through Carnivora, but continue on up through Insectivora and primates. Progressive central nervous system and brain development and enlargement—the unwavering evolutionary result that has followed placentation changes in mammals—is accentuated in primates. Satisfaction of the progressively larger oxygen demand of the developing mammalian egg in the latest stages of gestation has resulted in morphological and functional division of the egg into the larger-brained fetus on the one hand and its respiratory organ, the placenta, on the other. This division of the egg at term also modifies uterine delivery mechanics. The latest results of these two paths of evolutionary change in primates, the primary one—delivery mechanics—and the secondary one—egg respiration and nutrition—are complete fusion of the Muellerian ducts in formation of the primate uterus and the development of a very vascular and deciduous endometrial surface. The deeper endometrial invasion of the chorion into the lumens of endometrial blood vessels forming villi, thereby establishing greater respiration within a smaller uterine chamber surface area, has necessarily added endometrial blood-vessel clamping physiology to the mechanical separating and emptying functions of the primate uterus.

This endometrial blood-clamping physiology has developed as a part of the growth and enlargement of uterine musculature, further modifying, in upper primates, the basic mechanical physiology that moves the egg through the duct. It is evolutional addition of still more specialized mechanical physiology, change that affects uterine physiology of the entire primate reproductive (estrous) cycle, change in physiology that injures the maternal blood circulation in surface endometrium and produces menstruation in upper primates.

Wislocki (1929, 1930) has noted that primate placentas cannot be sharply classified on the basis of placental form, which can show the same variations in both primitive as well as more recent primate forms. He reported a "cleidoscopic series of changes in outer form of the placenta before it reaches the definitive form at the end of gestation." It is to be expected that primate uteri closely associated with the variations in placental form could also show complementary and supplementary variations in both primitive and recent primate forms. In infraprimate mammals the wide variations in degree of Muellerian fusion have been clearly classified: No fusion occurs at all resulting in two complete separate uteri and vaginae in monotremes and marsupials; a single vagina but two uteri are seen in most rodents; partial fusion, forming the bipartite uterus, is found in ruminants and carnivores; and a bicornuate uterus occurs in sheep. Even though Muellerian fusion may appear to be structurally complete in the formation of some primate uteri, the functional union is often incomplete. This functionally incomplete fusion is discernible more in the morphology and detachment of the mature placenta. The tubular uterus in infraprimates changed in stages of Muellerian fusion to the single-chambered primate organ.

The cleidoscopic changes in primate placental morphology that defy classification efforts are expansions of microscopic structuralization. Microscopic structure of mammalian placentae in infraprimate orders is more amenable to classification as Grosser (1909) has demonstrated. Enough extinction in "dead end" lineages has left sufficiently wide gaps in existing lineages to allow the more convenient grouping into ascending orders. The mammalian placenta, however, in its formation during gestation, in most instances goes through several earlier evolutionary stages before reaching its mature structure and form. Although a few forms of placentation, such as that seen in the rabbit, show two types of placental structure and function at maturity, most mammalian placentas mature along the course of their formation to a single type. Mammalian lineages in the primate order, for the most part, being more recent and more closely related, with fewer and narrower separating gaps, show much less variation in placentation. The evolutionary record, more fully revealed its gradual steps and the closer relationship of variation in primate branches, can appear confusing. Variations seen in primate placentation are variations in both microscopic structure and gross morphology that are secondary to progressive fusion of the Muellerian ducts. Although variation in gross morphology of the primate placenta must be studied and also considered, the differences in microscopic structure and implantation are much fewer, making it easier to follow the evolutionary path, which leads to a better understanding of placental morphology in the primate as well as menstrual bleeding in the primate estrous cycle.

Placentation studies indicate the ranking of lower catarrhine monkeys between platyrrhine and upper catarrhine monkeys, and upper catarrhine monkeys between lower catarrhines and the great apes, including the human lineage. This ranking of primates according to the advancement of their placentation is in full agreement with the ranking of mammals in general according to the developed level of their central nervous systems.

The earliness in gestation and the degree to which mesodermal cores push into the trophoblastic syncytial covering of the trabeculae to produce mesodermal respiratory villi that protrude into a labyrinth that enlarges into an intervillous "space" is gradual in the ascending scale of upper primates. Endometrial vessels are penetrated earlier in placentation of the great apes and the human than in the placentation of other catarrhini, to increase the respiratory functions of placentation. Villi also are produced earlier in placentation of the great apes and humans when compared with placentation of other catarrhini. Villi provide more respiratory surface than trabeculae. Less area of smooth chorion forms, as earlier and deeper endometrial penetration of a still smaller egg is achieved in upper primates. The great apes and humans appear to have advanced farthest past structural limits on the evolutionary path of reduction in ovarian egg size. Their particulate eggs contain only enough yolk to carry cleavage to the formation of a hollow blastocyst no larger than the tiny egg itself. The tiny blastocyst erodes the endometrial epithelium forming a small porelike opening, an entrance, through which it enters and burrows into the endometrial stroma by amoeboid movement. The entrance pore in the endometrial epithelium is then sealed over by an operculum of coagulum completely burying the egg beneath the endometrial surface membrane. The histotrophic stage of nutrition, so lengthy in infraprimate orders and briefer in lower primates, is so brief as to be hardly existent in upper primates; the early forming trophoblastic villi form lucunae, penetrate endometrial capillaries immediately, and establish hemotrophic egg nutrition and respiration as quickly as is structurally possible. From loss of yolk and egg size the surface-attaching primate egg progresses to a particulate, endometrial-burrowing egg that, in upper primates, attaches within rather than on the surface of the endometrium. By burying and sealing itself beneath the surface epithelium of the endometrium, the egg particle establishes a capsular, decidual layer above it. It is the presence and effect of this capsularis that distinguishes placentation of humans and great apes from catarrhine monkeys; a single discoid placenta and embryo form completely intramembranously rather than partially within the uterine cavity.

All mammalian eggs begin development as a tiny sphere and continue to enlarge and grow in that form. The geometrical shape and configuration of the uterine chamber in placental mammals alter egg shape, the single-chambered uterus of primates commuting the egg's spherical shape to an ovoid one. If the egg is laid externally, it retains its spherical shape; if its growth and development are within the cylindrical cavity of a uterine duct or horn, it becomes elongated accordingly. The chamber walls, however, composed of smooth musculature, also grow, enlarge, and distend from both physical pressure and endocrine stimulation from the egg. The shape of the uterine chamber formed by the Muellrian ducts varies from a communication between two tubular horns in infraprimate mammals to the inverted cone-shaped chamber in primates. With more muscular development and growth, a dome or fundus is developed in the uterus of upper primates. With fundic development in the uterus of higher primates, the cone shape of the chamber enlarges more spherically during gestation as it accommodates egg enlargement. The fusion of the Muellerian ducts accomplishes a reinforcement of the longitudinal uterine muscle fibers that provides a slinglike mechanical force during parturition of a large, weighty, and bulky upper primate fetus. Although the anatomical shape of the uterine chamber of upper primates is an inverted cone at the beginning of the reproductive cycle, as gestation progresses its functioning shape changes slightly with growth of its muscular walls. The slight change is critical. Thinning and distending musculature grows in such proportions that more thickened anterior and posterior walls are formed about a uterine chamber that becomes flat and obliterated when the organ completely empties at parturition.

Uterine contraction and involution at parturition in the primate is more complex than the simpler peristalticlike waves in horns of infraprimate uteri. Complexity of contraction and shrinkage of the circular and longitudinal muscle fibers of the uterine chamber in primates during parturition centers about geometry of muscular-directed pressure on the egg. This geometry results from evolutionary changes in muscle-layer morphology of the uterus. The morphological changes account for development of mechanics that detach the single egg, which is deeply attached to its endometrial bed lining and lying within the single uterine chamber. Cervical function and structure are required in this mechanical physiology just as cervical function is required in parturition of single-fetus ungulates. Fetal respiration in ungulates has to be maintained through the final stage of its hatching and delivery from a comparatively enormous egg superficially attached over an extended surface area. The egg is mostly extramembranous, lying mostly within the uterine lumen rather than within the endometrium, which is the condition that characterizes placentation of the highest-ranking primates.

Respiration likewise must be maintained in the primate fetus during its hatching and delivery from an egg, which although mechanically attached superficially, is much deeper and intimately attached functionally within a more confined surface of the endometrium. The upper primate egg develops completely intramembranously and from the standpoint of development does not lie in the uterine lumen or cavity. The egg chorion develops so intimately attached to the endometrium that it cannot be separated from it even partially. The upper primate egg is not separated at delivery. Only the endometrial surface membrane containing the egg is detached and separated with the attached egg still contained within. It is only the vascular connection of maternal tissue (the intervillous sinus that develops from endometrial surface capillaries) to the deciduous surface membrane that is severed at the end of gestation and the estrous cycle.

The cervical function and muscular delivery mechanics of primate uteri are determined by the shape and depth of egg attachment to the endometrium. The functions become established at the beginning of the reproductive cycle when the egg is produced at ovulation. With deeper egg invasion and its endometrial vascular communication established at the beginning of the reproductive cycle in upper primates, the uterine parturition mechanics include vascular staunching and bleeding control that characterizes the menstrual cycle of these primates. This is in contrast to the egg in lower primates and infraprimate mammals that does not immediately invade the endometrium during fixation and attachment. Complete removal of the unused and residual endometrial surface in the estrous cycle and postpartum period of these lower forms is by phagocytic absorption; bleeding and staunching physiology is unnecessary.

Although either the bilobed placenta or the single-lobed one is usually characteristic of a particular primate form, nevertheless both placental forms overlap in various primate lineages. Even in a single primate species both variations of placentation and placental form may be found. Many placental anomalies in the human are placental forms that are more consistent and normal in infrahuman primates. Often these human placental anomalies are aberrations in egg growth and development that stem from variation in depth to which the egg in that particular pregnancy happened to attach to the endometrium.

The lobes of the placenta in lower, more primitive primate forms are separated by smooth avillous chorion, corresponding to the bare poles of the carnivoral egg previously described. The extent of smooth chorionic surface in lower primate placentas indicates the depth of early egg invasion into the endometrium. Chorionic surface in contact with the uterine lumen and which does not engage in trophoblastic invasion remains smooth and bare of trophoblast. In both Carnivora and lower primates the invasion depth of the egg may be reckoned relative to its diameter at attachment. The carnivoral egg is larger from imbibed secretions before beginning its invasive endometrial attachment. The smaller egg in lower primates begins invasive attachment on the surface of the endometrium covering one wall (anterior or posterior) of the uterus. This primary site of invasion forms the primary lobe of the placenta. The particulate-sized egg does not completely enter the endometrial surface. The tiny egg sits on top of the endometrium invading and attaching from its lower pole in contact with the endometrial surface while expanding its upper domelike, smooth surface into the uterine lumen and may begin invasion and attachment to the endometrial surface of the opposite uterine wall, forming a secondary placental lobe. When two separate lobes are formed they may also be secondarily lobulated. The umbilical cord in lower primates is almost always inserted into the margin of the primary lobe with vessels from a wide umbilical pedicle passing between the sulci of the primary lobules across the smooth area of the chorion to the secondary lobe. The vessel pattern results from the early dominance of respiratory mesoderm growth in the primary lobe of a tiny, partially embedded egg.

The endometrium in all mammals possesses absorptive functions. Resorption of products of conception or maternal placental tissue (contained within the deciduous endometrial surface) by phagocytosis or otherwise is as important a function in the postpartum period of the reproductive cycle of single-fetus mammals as it is in the antepartum period of litter-bearing forms. A litter must be protected from the loss of one or more of its individuals, and the endometrium absorbs the conception products or mummifies them until parturition. In single-fetus mammals loss of the single conception results in its mechanical expulsion, the mechanical emptying of the uterine cavity, completing termination of the cycle by endometrial phagocytosis and the institution of a new cycle. The same occurs in gestation of litter-bearing mammals when the last individual in the litter succumbs either before or after parturition. These are the functional conditions of the estrous cycle in which the reproductive cycle is ended by the uterine-emptying physiology that empties the uterine cavity of its maternal placental tissue when it contains no attached fetal placenta. In the hemotrophic cycle of upper primates that portion of the endometrium produced to house the pregnancy is subjected to the emptying squeeze of the uterus. Separation and delivery of the barely formed membrane add hemorrhage to the leukocytic infiltration that removes the damaged membrane debris.

The squeeze of parturition physiology at the end of the reproductive cycle staunches the arterial circulation to the deciduous endometrial surface, whether it contains the attached egg placenta when the cycle ends at parturition, or whether it contains no egg at all when the cycle ends in an estrous cycle. Parturition ischemia of the placental site is parturition ischemia of the endometrial layer or level at the depth at which trophoblast invades. From the geometric and mechanical standpoint, in all placental mammals the egg invades only the surface of the endometrium at the site of its fixation and attachment, leaving most of the egg mechanically protruding into and positioned within the uterine lumen. The richer capillary blood circulation found in the deeper layers of the cellular endometrium of lower primates and infraprimates is developed closer to the endometrial surface in the higher-ranking, larger-brained menstruating primates. Although the egg in these forms remains completely within the endometrium from the standpoint of development, from the standpoint of mechanical force exerted by uterine musculature the egg, tremendously enlarging by virtue of its embryo or fetus, fills and remains within the uterine cavity. The endometrium, therefore, has evolved structurally and functionally into two membranes in the primate—a deciduous surface containing the egg and a nondeciduous deeper layer attached to the inner muscle layer of the uterus. The deeper one contains the blood supply to both membranes. The two endometrial membranes function differently and mechanically separate under the mechanical forces of parturition. The blood supply to the deeper nondeciduous membrane is not damaged, but the blood supply to the surface layer is staunched or occluded by the mechanical squeeze of parturition. The inner muscular layer of the uterus that squeezes and separates the two endometrial membranes also concomitantly staunches the blood supply to the separating surface membrane. The structural differences of the two membranes center about the formation of two different arterial blood vessel systems: a system of spiral arterioles that supply the decidual surface membrane and a system of straight arterioles that supply the deeper attached adeciduate membrane. The mechanical squeeze of parturition that empties the uterus synchronously sphincterizes the bases of the spiral arterioles clamping off all endometrial surface blood circulation. When the surface layer is separated and passed by the mechanical parturition squeeze applied early in the reproductive cycle, when the layer contains no egg or placenta, the function is necessarily a lethal strangulation of the biologically dismissed tissue layer (the egg's useless endometrial bed). The condemned superficial decidual endometrial layer previously was destined to comprise a maternal placenta. Return of endometrial blood circulation of the new cycle to the dead, discounted, but still unseparated endometrial layer of the old cycle results in menstruation, i.e., endometrial phagocytosis, bleeding, clot formation, clot liquefying, cellular slough, autolysis, and separation and passage of the superficial placental layer of the endometrium. The surge of fresh arterial blood into the network of dead capillaries passes through the necrotic capillary walls, forming stromal hematomas and hemorrhages. The dead, autolysing endometrial surface is separated by undermining surface epithelial growth from the glands of the new cycle that covers over the viable, deeper endometrial layer. Perhaps hemorrhage beneath the dead layer of tissue may also be part of nature's way of separating the adherent slough, i.e., the cleansing function theorized by the ancient Greeks. Some hemorrhage may be necessary in sequestering the dead endometrial layer, and incarcerated blood from hemorrhage may give volume to the surface, enabling the uterus to mechanically pass or deliver it as a volume.

Accompanying and resulting from an enlarging brain in the human is stereoscopic vision; upright posture; a greater use of the forelimbs, face, and mouth parts; and an enlarging chest containing larger lungs and heart. Regulation of body heat in such a large-brained, oxygen-consuming mammal as the human has resulted in loss of body hair to expedite body heat loss. Coitus and fertilization show little modification from that of mammals in general. Incubation of the egg also is little different from that of other placental mammals. Hatching and delivery of the egg at parturition has become different in various mammals, but much more different in primates and humans, with the largest developed brains and central nervous systems. Safe severance of the most intimate respiratory attachment of the egg from the mother's blood circulation incurred the development of the vascular staunching mechanics in parturition and the menstrual cycle just described. Details of this mechanical physiology in the human are considered and illustrated in the chapters that follow.

Summary

Modern primates reveal gradual steps in the evolution of menstruation within their estrous cycles. Because primates are closely related, the steps are short, and the numerous small variations in primate reproduction make the evolutionary path more difficult to discern. Nevertheless, when reproduction in larger groupings of primates are compared, extension of the path of infraprimate mammals can be determined.

Lower primates have an estrous or reproductive cycle similar to the estrous or reproductive cycle of infraprimate mammals. The egg utilizes its stored yolk for cleavage only, then begins to absorb uterine glandular secretion and the "pabulum" of destroyed endometrial epithelium and stroma without damage to endometrial blood vessels. Although very brief, this period of egg nutrition leaves their abbreviated reproductive cycle, their estrous cycle, essentially unchanged from infraprimate cycles.

In upper primates, however, the shortening histotrophic stage of egg nutrition has become so brief that the egg establishes hematrophic nutrition during its implantation and attachment to the endometrium. This changes the character of the abbreviated cycle by adding the factor of hemorrhage to endometrial leukocytic absorption.

Physiology of the menstrual cycle that has evolved in upper primates is physiology that controls endometrial hemorrhage in their reproduction.

Part Two

Physiologic Mechanics of the Human Uterus

IV
Function of the Human Uterus before Labor

In the past, some textbook authors have expressed the view that the uterus has two functions, namely, menstruation and childbearing, with the first being a preparation for the second (Crossen and Crossen, 1944). Although that relationship may have seemed apparent in the past, no evidence surfaced later that proved menstruation to be a preparatory function. Indeed, failure to comprehend the significance of menstruation introduced uncertainty into the results of all direct studies of the phenomenon. Some current textbook authors (Herbst et al., 1992) make note of the fact that relatively little research on the mechanism of menstruation has been carried out since the works of Bartelmez (1933) and Markee (1940). In the last half-century, gynecologists have been concerned with managing menstrual disorders, and obstetricians have been concerned with childbirth. These divided efforts by clinicians to distinguish between uterine function associated with menstruation from that of childbearing have been unsuccessful in shedding additional light on the problem of explaining menstruation. Regarding the physiology associated with menstruation as different from physiology of childbirth has contributed to failure in recognizing identical physiology in the two uterine performances. Although a full understanding of menstruation is the goal of efforts here, identifying its physiology as childbirth physiology brings to question the physiology of other puzzling uterine performances. Besides menstruation there are the haunting enigmas of dysmenorrhea and the phase of severe endometrial ischemia preceding menstrual bleeding that have defied explanation. These additional unexplained phenomena in the uterus raise the further question of how well the role of the uterus in reproduction is clinically understood. Much of the childbearing performance of the uterus may not be as clear as it may seem to some; our present understanding of uterine function is fragmentary and weak. The uterus evolved as function and structure that delivers and hatches the huge mammalian egg at the end of gestation. No one can doubt that childbearing is the major role of the human uterus and is the dearest and best interest of clinicians. A full clarification of childbirth physiology must be established before an insight into other mysterious uterine behavior can be gained. Accordingly, interest in understanding menstruation at this time must necessarily revert to efforts directed toward deriving a better understanding of the role of the entire uterus in childbirth, even though present interests in obstetrics and gynecology seem to focus more on functions of the ovary and oviduct.

Childbearing, in its final analysis, is a mechanical accomplishment of living tissue, and obstetricians in the past devoted considerable attention to the study of childbearing mechanics. Subsequently, interest has been attracted to the splendid accomplishments in reproductive endocrinology, including the role of ovarian endocrines in uterine function. Though the marvelous results of these fascinating studies have greatly advanced our understanding of the endocrine ingredients of reproductive physiology, mysteries of uterine function are left still unsolved, and our understanding of uterine mechanics remains incomplete. Here, attention is returned to the study of childbirth mechanics and concentrated on the areas where previous studies left off. The terminology used in times past referred to the "powers," the "passage," and the "passenger." Studies of the passage were complete, for the most part, in the classification of different types of human bony pelvises. The powers and the passenger, or the uterus and its contents during delivery, are to be considered in further detail.

Power, or mechanical force exerted by living tissue, is provided by chemical change within its many contractile units which, in the case of the uterus, are smooth muscle cells. Shape, the morphology and geometrical configuration of contracting tissue attached to the immovable bony pelvis, is critical in determining the direction and intensity of the mechanical force it produces. Morphology of uterine contents also determines the direction of mechanical force produced by the uterus during labor. Without the presence of a firm fetal mass acting as a fulcrum or wedge, the cervix is not pulled and diverted, or driven, mechanically to its completely open state; structure of the uterus, as well as structure of its contents, is also critical in continuing blood circulation to the contracting uterine tissue while it contracts. Although a constant blood circulation to uterine musculature during contraction is critical in reproduction, its extent in the human uterus is nothing like the magnitude and extent of blood circulation required in the endometrium by the respiration of its contents during pregnancy. The relationship of structure and function in the uterus is such that function of the contracting uterus is determined by its own anatomical geometry and geometry of its contents.

Figure 4.1 represents the solid or formed egg contents of the human uterus delivered at term. The delivered contents consist of a detached, ruptured, and collapsed maternal membrane, which is the surface layer of endometrium or upper portion of decidua functionalis including the decidua compacta. The ruptured membrane, separated from its lower functionalis and basalis that remain as the lining of the uterine cavity, in turn contains the hatched or opened human egg emptied of all its free fluid and with its solid contents (fetus, cord, and placenta) still intact. Although the uterine membrane is present, it is hardly visible or detectable, being greatly overshadowed by the morphology of its contents, the solid structures of the egg consisting of fetus, cord, placenta, amniotic and chorionic membranes. The liquid structure, the amniotic fluid, now released and lost, only moments previously, provided a critical and vital mechanical support for the function of these structures during their encapsulation within the uterus. The conversion of uterine contents from their encapsulated and contained state (package) to their hatched, delivered, and free, unpackaged state is the marvelous accomplishment of uterine muscle contraction. The contraction is intricate and orderly. In preceding chapters it was pointed out that in the human and most existing mammals the egg is first hatched and then delivered, reversing the premammalian order of first delivering or laying the egg prior to its hatching.

Figure 4.2 includes the structures in figure 4.1, the human egg at term within the uterus, before labor begins. The uterus contracts intermittently as it has done throughout pregnancy. With intact undelivered contents completely encapsulated within, contractions of the uterus apply mechanical force or pressure equally over the peripherally attached surface of the decidua capsule. This maternal membrane is inseparably fused with the outer surface of the outer egg membrane (chorion) contained within. The egg membranes are also structured and fused as one mechanically closed membrane (amnion) enclosed within another mechanically closed membrane (chorion) that is enclosed or encased within the endometrial membrane (fig. 4.3). The innermost membranous enclosure, the amnion, confines both liquid and solid egg contents. It is sealed within the chorion, which is the intermediate membrane that is fused to and within the decidua compacta that forms the outer covering of the described egg contents in figure 4.1. The chorion is so firmly attached and locked within this superficial layer of the endometrium that the human egg cannot be sepa-

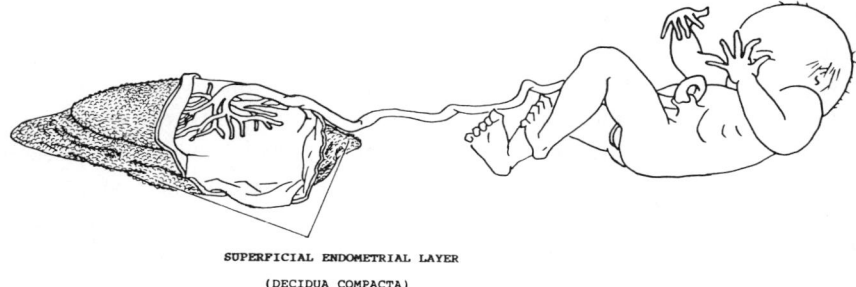

SUPERFICIAL ENDOMETRIAL LAYER
(DECIDUA COMPACTA)

Fig. 4.1. Representation of the solid, or formed, egg contents of the human uterus delivered at term. The delivered contents consist of a detached, ruptured, and collapsed maternal membrane, the surface layer of the endometrium, designated decidua compacta. The ruptured endometrial membrane in turn contains the hatched or opened human egg emptied of all its solids and free fluid. Although the uterine membrane is present at delivery, it is hardly visible or detectable, being greatly overshadowed by the size and morphology of its contents—the fetus, cord, and placental membrane.

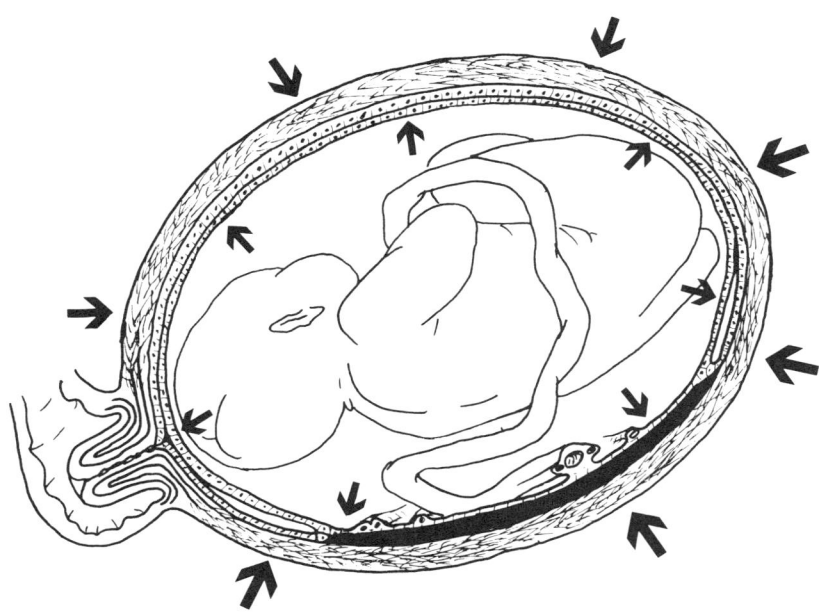

Fig. 4.2. The human egg at term within the uterus before labor begins. The structures in figure 4.1 are shown intact. The outside arrows indicate the mechanical pressure exerted by the uterus, and the inside arrows indicate the counterforce exerted by the amniotic fluid. The two pressure forces constitute a single "splinting" force for the placenta.

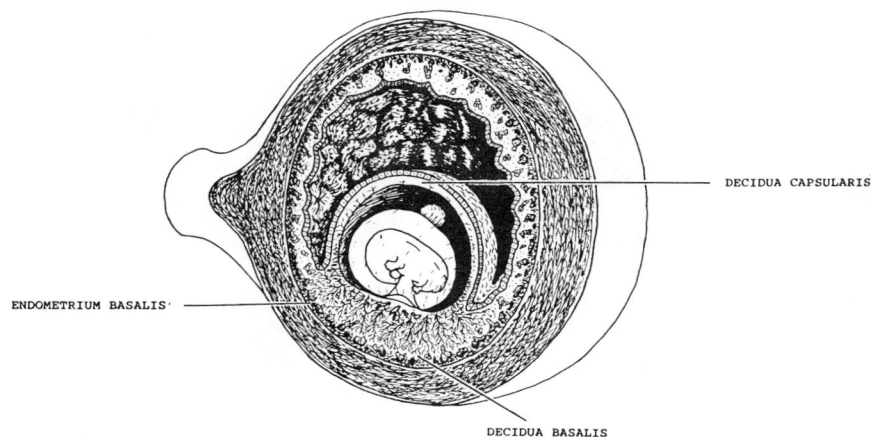

Fig. 4.3. Diagram of the human uterus in the third month of pregnancy demonstrating the formation and relationship of fetal and maternal membranes. The specimen has been sliced to the left of the midline cutting through the chorion and missing the amnion and its enclosed embryo. The three fluid compartments are shown; the amniotic sac, within the chorionic sac, within the fluid-filled uterine lumen. The patent internal os at this stage is shown but not labeled.

rated and delivered from this membrane in either its hatched or its unopened form. The uterus must invariably detach and deliver this encapsulating layer of the superficial endometrium as the means of accomplishing mechanical delivery of the egg.

The amniotic fluid is thus confined within the uterus by three different membrane layers fused to mechanically form a single reinforced membrane. The function of this three-layered membrane is vital to the survival of both mother and infant during uterine contraction, as both are protected from the effects of the increased mechanical pressure exerted on the peripheral surface of the egg. Typically, the uterus during labor mechanically opens that portion of the three-layered membrane lying just inside and attached to its dilating cervix and then expresses or drains forewater amniotic fluid followed in sequence by the fetal mass, the remainder of the amniotic fluid or hindwater, the placenta, and finally, the remainder of the extraplacental three-layered membrane. An opening or rupture of this membrane in another area before delivery that permits amniotic fluid to be expressed or injected into the venous circulation of the uterus can be fatal to the mother. Before labor, an opening in this membrane that directs the draining amniotic fluid to the cervical exit slightly reduces the volume and surface area of the uterine cavity. If intraamniotic infection does not occur, premature labor is likely. As long as the triple-layered membrane remains intact before labor, counterpressure of the contained amniotic fluid directed outward against the contracting uterine wall ensures sufficient volume or space to accommodate the growing and enlarging fetus and sufficient functioning placental surface to effect respiratory and metabolic exchange. The contents of the uterus before labor, the human egg with its intact membranes encapsulating the amniotic fluid, are incompressible. The incompressibility of the amniotic fluid provides complete protection and fully shields the submerged fetus, cord, and placenta from any mechanical pressure the contracting uterus applies to the encapsulated endometrial membrane. The solid egg contents are tissues that also consist mostly of fluid and mechanically can be regarded as a submerged system of membranes with equal fluid pressure simultaneously applied on both sides of every membrane surface. There are no compressible states of either liquid or formed elements within the egg as long as the three-layered membrane covering remains intact. The amniotic fluid, completely confined, thus provides full mechanical support and protection of egg contents. The respiratory gases are held chemically bound in a solid state for chemical exchange in tissue, blood, and the placenta, and never convert to their free, compressible, gaseous state in living egg tissues during their intrauterine encapsulation.

The source of egg oxygen and the vehicle of departure for the egg's carbon dioxide waste is the enormous endometrial blood circulation of the uterus. Through this extensive maternal circulation, the respiratory gases of the egg are conveyed to and from the mother's lungs. By the same means other metabolic waste of the egg is conveyed to the mother's kidneys. Within the confining egg membranes at term, the fetus urinates into its amniotic fluid, which it reabsorbs through its skin and swallows into its alimentary canal. The metabolic waste thus reabsorbed into the fetal circulation is returned to the mother's circulation via the placenta and distributed to her kidneys. The volume of urine that the fetus urinates adds to the volume of amniotic fluid and contributes to the counterpressure the amniotic fluid exerts against the contracting uterine musculature. The absorption of amniotic water through fetal skin and alimentary canal and the addition of urine to the amniotic space provide a circulation that normally at least partly determines the volume and pressure of the amniotic fluid. Undissolvable constituents of the amniotic fluid collect as vernix and meconium on the collecting surfaces of the fetus.

The secretion of amniotic water by the lining epithelium of the inner surface of the amnion may be comparatively constant. Thus the living fetus *in utero* functions to provide its own growing space through adjustment of amniotic fluid volume and pressure. In cases of fetal renal agenesis, reduction in amniotic fluid volume may occur, resulting in a difficult "dry labor." In cases of congenital intestinal atresia, polyhydranios may occur. A normally increasing volume of amniotic fluid increases space for fetal growth and enlargement, and does so by increasing amniotic fluid pressure outward against the intermittently contracting uterine musculature. During pregnancy the intact triple-layered membrane conveys this pressure to the contracting musculature that responds in turn by distending, thinning, and expanding its walls through the method of proliferation, growth, and enlargement of its individual smooth muscle cells. The pattern of structure in the outer two layers of uterine musculature does not change markedly during gestation; only the inner surface of the muscular wall of the organ responds, increasing to accommodate the enlarging volume of cavity contents. The other enlarging smooth muscle layers lengthen by expanding and distending about the enlarging uterine blood vessels that supply the growing and enlarging contents of the uterine cavity. The smooth muscle cells are structured to contract parallel or oblique to, rather than across, the lumen of the vascular (venous) channels; otherwise their contraction would be a sphincter effect and close rather than pull open the vascular lumens. The placental blood circulation is established on the venous side of the endometrial capillary circulation; consequently, the venous circulation is in greatest risk during contraction of the uterus. The venous channels through the uterine musculature are tortuous and circuitous but do not form a random plexus. The channels progressively enlarge in diameter throughout their course to the hilus, as all large venous channels do that do not form a portal system.

With the muscular wall of the term uterus only a half-centimeter in thickness, the large blood vessels supplying the pregnant uterus occupy a large portion of its wall, especially at the organ's hilus. An intermittent contraction of the pregnant uterus begins as a generalized tightening of the inner layer of uterine musculature that lies nearest to and is attached to the endometrium. This layer is penetrated by the finest arterial and venous channels that provide blood circulation to the "intervillous spaces" of the placenta. These fine arterioles terminate in and the venules originate in the capillaries of the decidua compacta that at term have developed into the intervillous "space" (fig. 4.4). These are the fine blood vessels that are severed when the placenta is delivered (fig. 4.5). The contraction of uterine musculature begins about the fine venules in this inner muscular layer and follows the course of the venous channels to the hilus of the organ. Because the incompressible egg within presents a counteracting force, a "milking" effect is produced in the larger venous channels. Uterine contractions drive or milk the venous blood flow away from rather than toward the placenta. As the muscular tightening reaches the surface or peripheral layers a blanching may be detectable. Some have described "waves" of contractions.

Although uterine blood flow may diminish during a uterine contraction, placental circulation is maintained throughout the contraction unaffected. Arterial blood pressure drives the maternal blood through the intervillous "space" of the placenta which, like the fetus and its umbilical cord, is completely protected from the effect of mechanical pressure of uterine contractions. Placental blood circulation is preserved during uterine contraction by the splinting effect of the sealed amniotic fluid.

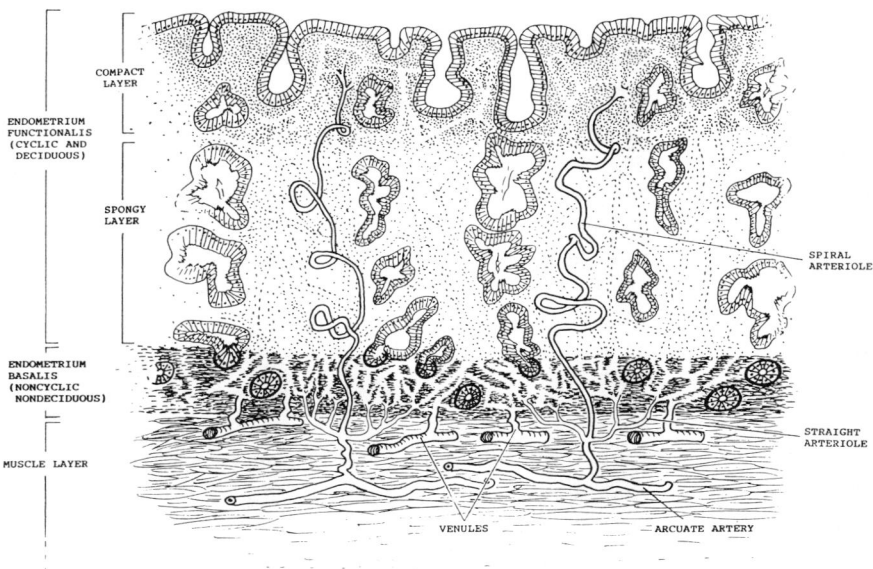

Fig. 4.4. Main structures of the human endometrium prior to egg implantation as seen in tissue sections prepared for microscopic study. The endometrial surface or compacta is vascularized by the spiral arterioles, which are terminal arterioles. There are no corresponding spiral veins. All capillaries empty into the straight veins of the endometrium basalis.

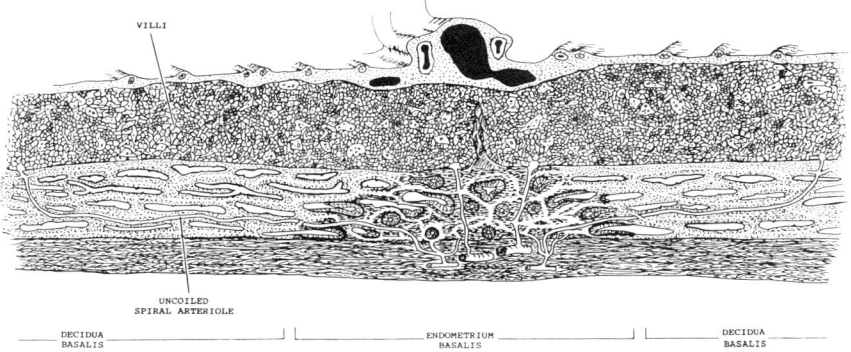

Fig. 4.5. Diagram of a section through the intact human placenta near its junction with the umbilical cord. The organ forms and functions within the capillaries of the stretched and expanded deciduous endometrial surface. It is held and braced in this strategic and precarious location by "splinting" effect of fluid counterpressure.

V
Function of the Human Uterus during Labor

The relation of the uterus to its contents represented in figure 4.2 is mutual growth and enlargement. As the egg acquires volume and enlarges so does the uterus enlarge in accommodation. However, the larger the spherical-shaped egg becomes the more absolute volume it acquires in proportion to incremental increase in its surface area. The uterus, in accommodating to the greater egg volume, does so with proportionally less inner surface area. Although uterine walls thin and distend during gestation, their individual muscle cells lengthen and thicken about progressively enlarging blood vessels that must be kept patent to maintain a functioning circulation, while the uterus carries out its reproductive role. It is known that most of this uterine growth during gestation is induced by endocrines from the organ's contents. The growing egg within the uterus produces and releases the same estrogenic hormone substances that the ovary produces at the beginning of the reproductive cycle. These hormones stimulate uterine growth. Uterine musculature receives growth or estrogenic stimulus at the beginning of the human reproductive cycle from the ovary only, and only through the general circulation. After implantation this further stimulus of uterine growth from uterine contents is both directly from forming placental tissue within the endometrium, and indirectly from additional but relatively diminishing estrogenic and other hormones being released by the ovary into the general circulation. As the uterus enlarges, its further growth requires still larger quantities of estrogenic substances, so that as the egg grows it releases increasingly larger amounts of estrogens sufficient to further incur steady growth in the continuously enlarging uterine musculature. The rising level of circulating estrogens and progestins may thus be looked upon as a representation of the amount of growth stimulus necessary to stay ahead of uterine growth. Once this increasing growth stimulus to the uterus falters or lessens its crescendo, imminent mechanical emptying of the uterus becomes a threat. The emptying functions of the uterus cease to be latent or postponed (see chaps. 2 and 3), involution of the musculature, i.e., labor, may begin. Accompanying the increased production of estrogens is an increased production of progesterone by the egg. Progesterone, in contrast to estrogens, does not have prominent systemic effects. Its most noticeable effects are locally upon uterine musculature. It primes the enlarging mass of uterine musculature for involution or shrinkage, the opposite of growth. It sets up the involution or shrinkage physiology in the uterus that estrogens postpone or delay. Through involution of the enlarged uterine musculature, the uterus returns to its small, nonpregnant size after pregnancy, the approximate size it may maintain during the menstrual cycle. (See the works of A.I. Csapo, 1954 and 1979.)

At any particular moment uterine musculature may be said to exist in one of three states: (1) It may be in its functional phase of growing under estrogen stimulus, (2) it may be in its functional phase of involuting during labor and over the postpartum period, or (3) it may be resting in between changes from one function to the other. These three states of uterine musculature are illustrated in figure 5.1. When the estrogen and progesterone blood levels drop before labor begins, the enlarged uterine muscle mass ceases to grow and rests briefly from growth, even though intermittent contractions continue. After involution is complete during the early nursing period before menses return, the musculature of the small uterus also rests. In early childhood and after the menopause, when uterine musculature is relatively free from endocrine stimulation, it may rest in its small, involuted state.

There are two different types of uterine contraction: (1) intermittent contractions that the uterus prominently emphasizes and displays as it grows during pregnancy, and (2) the contraction of labor displayed as shrinkage and involution of the entire organ to its prepregnant size. Intermittent uterine contractions constantly occur throughout pregnancy, labor, and the postpartum period. Under the conditions represented in figure 4.2, before true labor begins, these intermittent contractions, or "squeezes" can be either painless Braxton Hicks or false labor contractions. They do, of course, usually progress into painful labor contractions during the involution of labor, and very painful contractions in the form of severe cramps in the postpartum period. Obviously both types of muscle fiber contraction can and do occur concomitantly in true labor. It is the second type of contraction, shrinkage and involution of muscle cells, that appears to change the character and intensity of intermittent contractions.

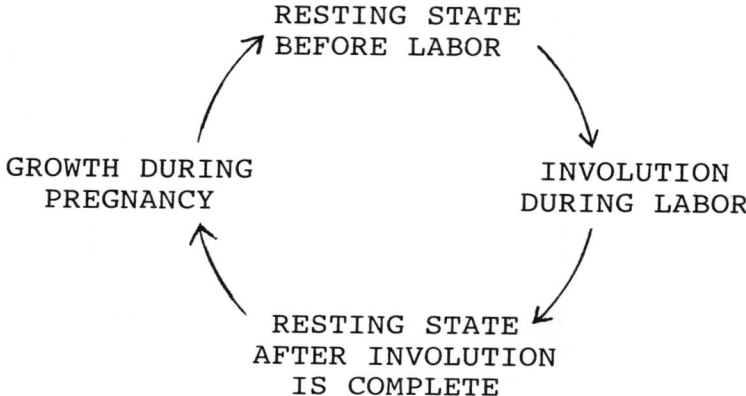

Fig. 5.1. Diagram indicating the two functional states of uterine musculature—growth and involution. In the cycle of their changes these two active states are separated by a state of rest that is variable in its length.

After an intermittent contraction the uterus rests, and each contracting muscle cell or fiber is usually considered to resume its former shape and length, i.e., the same size and length it had before the contraction. In its growth and enlarging phase during pregnancy, however, it should be considered that there is an incremental growth in size of the muscle fiber after the contraction, as growing uterine musculature accommodates increasing egg size. During the involution phase of the reproductive cycle the opposite occurs. The shrinking muscle fiber can be regarded as smaller in size and shorter after the intermittent contraction.

In the beginning of labor, involution is not generalized throughout uterine musculature like intermittent uterine contractions, but occurs first in certain layers and regions and gradually spreads to the remaining muscle layers in an orderly sequence. It is this progressive and orderly shrinkage and shortening of uterine muscle fibers after each intermittent contraction, spreading in steps and stages to specific areas within uterine musculature, that converts the painless Braxton Hicks and false labor contraction into true progressive labor contractions.

Before labor, generalized intermittent uterine contractions around the region of the uterine neck, or cervix, contract in a circular binding manner that results in an equal squeeze over the entire incompressible egg surface. This binding action at the cervix, and over the lower uterine segment when the cervix effaces, counteracts and neutralizes the squeeze in the remainder of the contracting musculature. This mechanical effect is represented in figure 5.2. Underneath this circular, constricting musculature, more longitudinally contracting muscle layers are continuous with cervical tissue. The uneffaced and undilated cervix may be regarded as a "knob" of deeper uterine muscle tissue protruding through the circular binding layer. This protruding knob of tissue containing a patent canal is a volume of uterine tissue that is converted to a surface during labor. The cervical knob disappears as it becomes the anterior inner surface of the uterine cavity. This performance is critical in opening and hatching the egg during normal labor.

Labor, or involution of musculature, begins in this muscle layer of the uterus lying intermediate between the outer circular binding layer and the inner layer directly underlying the placenta. The outermost layer of uterine musculature is anchored to the bony pelvis (figs. 5.3 and 5.4). The shrinkage and contraction of the intermediate muscle layer attached to or continuous with the cervix draws the cervical tissue over the incompressible egg when labor begins. The portion of uterine muscle lying immedi-

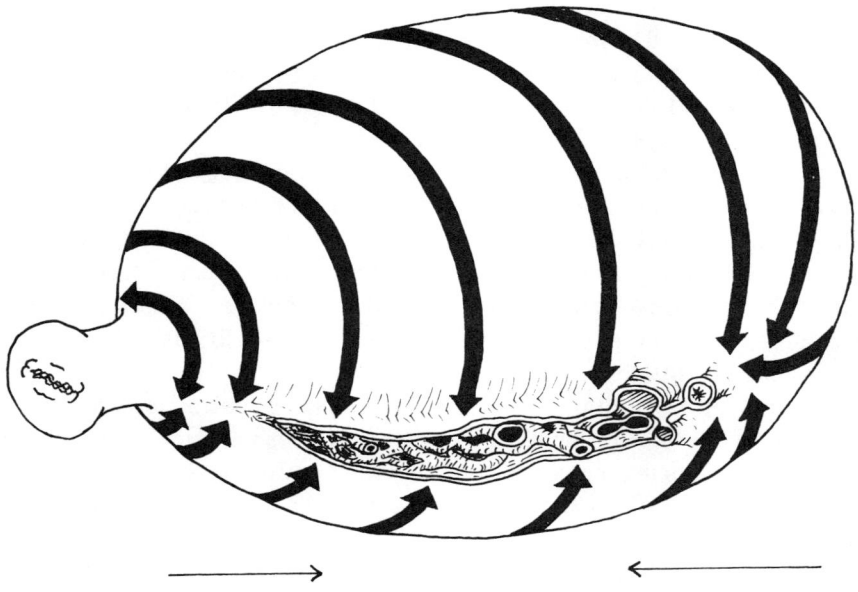

Fig. 5.2. Pregnant uterus at term isolated from its attachments showing direction and effect of the mechanical force exerted by an intermittent uterine contraction before onset of involution (labor). The effect is entirely within the venous channels of the muscular walls as long as the amniotic fluid is mechanically contained within the uterine cavity. The venous channels are compressed beginning at their origin in the subplacental muscular area and extending toward and into the uterine hilus. The egg and its contents are unaffected. The expulsive force of the contracting fundus is counterbalanced by similar contraction in the circular musculature around the base of the cervix.

ately beneath the placenta does not involute until the very last, converting the last intermittent, mechanical contraction to one that separates and delivers the placenta. It is the remaining extraplacental region of overlying uterine muscle, i.e., the inner surface of the muscle layers that cover the fetal area of the egg and is continuous with cervical tissue during labor, that begins involution first (figs. 5.5, 5.6, and 5.7).

The "layers" of uterine musculature referred to here are actually areas, or regions, of the musculature, rather than separate anatomic layers. The areas contract in unison intermittently, but one area may involute more rapidly than another; thus its position and spatial relations change during labor. The layer underlying the placental attachment necessarily tightens intermittently to brace and splint the placenta during an intermittent uterine contraction, but it does not shrink or involute during contraction in the first and most of the second stages of labor. It involutes and contracts last in the third labor stage when it accomplishes delivery of the placenta.

At the beginning of labor the mechanical relations represented in figure 4.2 are such that the closed internal cervical os can be considered to represent a point on the membranous surface of the intact egg coverings (fig. 5.8). This is an actual stellate structure on the membranes of some lower mammals, in which there is no decidua capsularis (see chap. 2). It is a fixed and sealed point in the human, supported and maintained by contraction in the counteracting circular and binding muscle layer described in figure 5.2. This sealed point is a covering of decidua capsularis over the internal cervical os formed when the enlarging egg fills the uterine cavity at the end of the fourth month of pregnancy (fig. 4.3). The decidua capsularis fuses with parietal decidua at this stage of pregnancy obliterating the uterine cavity, but the internal cervical os, held closed by sphincter action of the circular binding muscle layer, becomes a point sealed over only by decidua capsularis. The decidua parietalis surrounding the opening of the internal os, like all the other areas of the decidua parietalis, fuses with the decidua capsularis. As labor begins, the involuting uterine muscle fibers continuous with the cervix pull it open from the inside against the attached surface of the incompressible fluid-filled egg. The sealed point is pulled into an expanding circle, or ring. The rim of the ring consists of the two fused decidual surface layers. Stretched across the area of this expanding circular ring is the single three-layered membrane consisting of decidua capsularis containing the fused amnion and chorion. The cervical canal is widened and reduced in length as the cervix thins. Its mucous

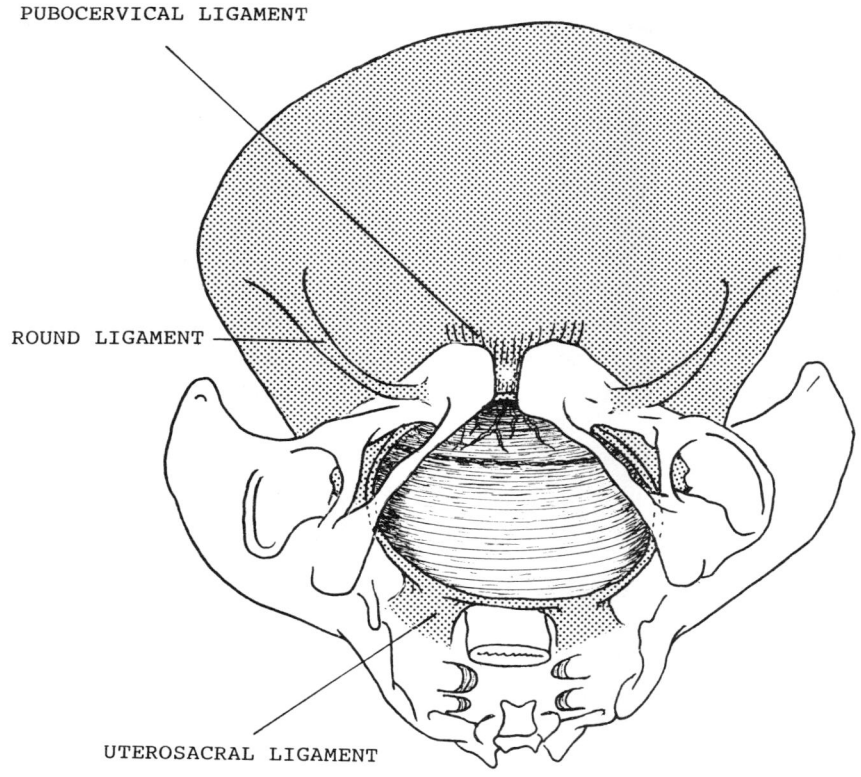

Fig. 5.3. The three ligaments anchoring the uterus to the bony pelvis during full cervical dilation. Vaginal tissues are not shown. Although the uterus in full cervical dilation remains an encapsulating muscular sac with its outside muscular layer pulling against its bony attachments, the inside muscular layer beneath the placenta and the venous channels remains distended and open respectively (a). The entrance and exit of functioning blood vessels at its hilus on each side functionally convert the muscular sac into a mechanical sling (b), composed of a single wide muscle band anchored anteriorly and posteriorly with blood vessels passing between its folds on both sides.

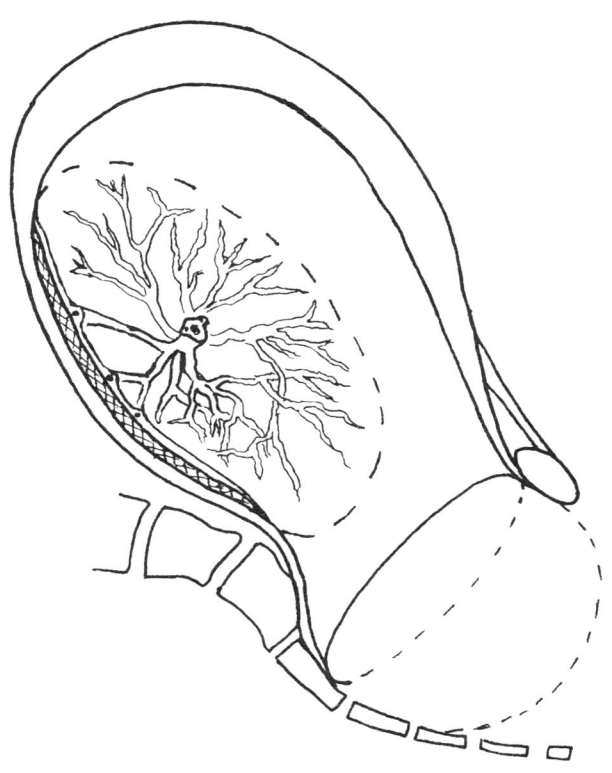

b

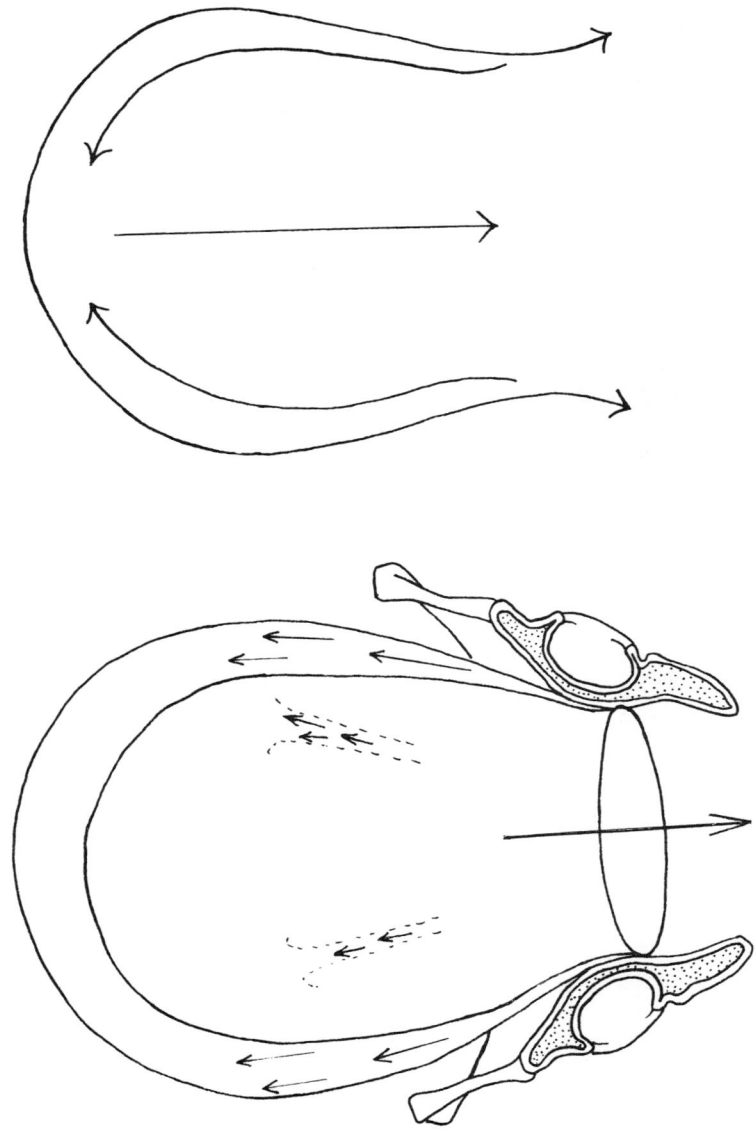

Fig. 5.4. Diagram of fully dilated cervix. When the cervix is fully dilated contraction of the uterus becomes shifted to its outside muscle layers that are attached to the bony pelvis. The uterus, in effect, becomes a contracting mechanical sling.

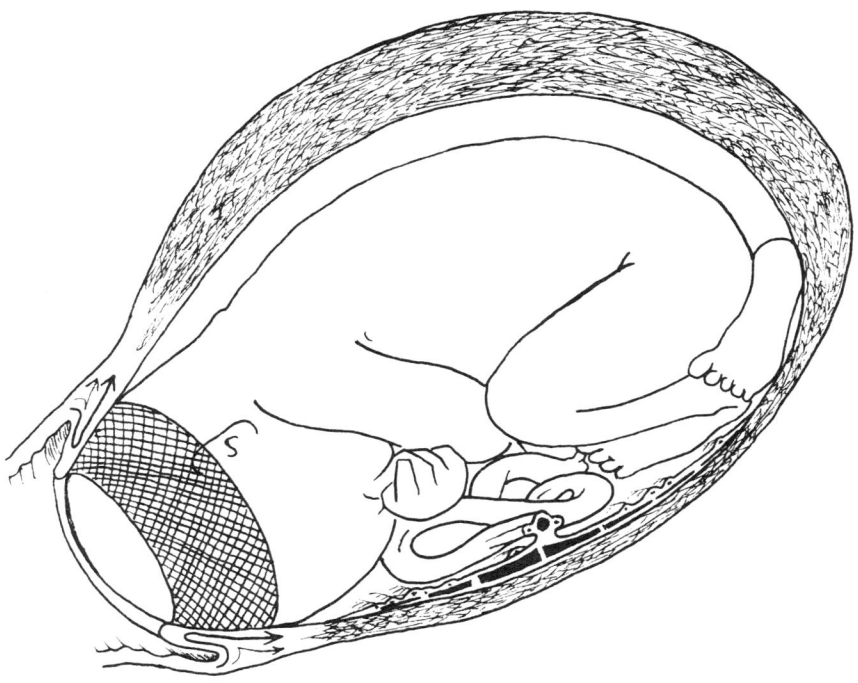

Fig. 5.5. Diagram of effacing cervix dilating and adding to inner surface area of the uterine cavity. As the uterine musculature overlying the fetus thickens, it maintains a constant volume of the uterine cavity and its contents. Only the presenting part of the fetus, the cross-hatched area shown, mechanically contacts the cervix and is subjected to mechanical pressure. The remainder of the fetus, cord, and placenta is protected from mechanical pressure by the splinting effects of amniotic fluid hindwater.

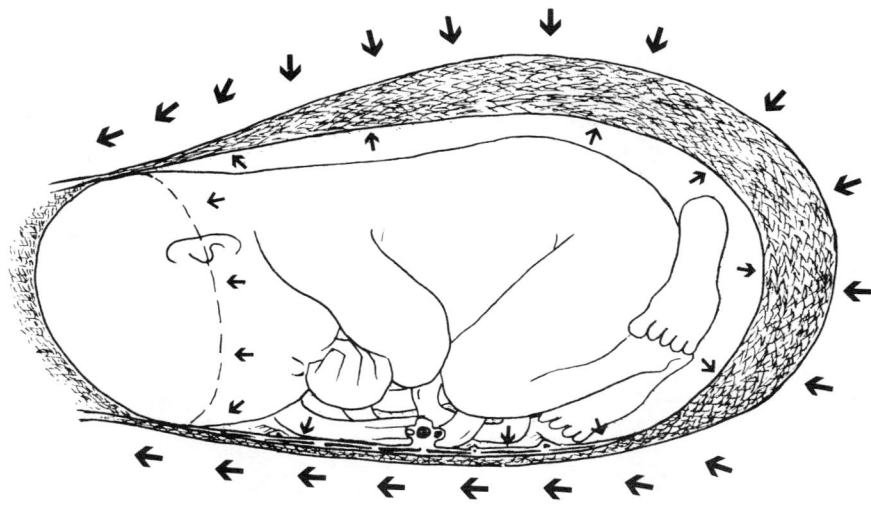

Fig. 5.6. The placenta maintained in its distended and functional state by the pressure of the hindwaters in the compartmentalized egg. Rupture of the membranes and release of amniotic fluid results in compartmentalization of both the egg and the uterine cavity. The compartments of the egg subjected to uterine contraction (fetus and amniotic fluid) become mechanical segments with each segment subjected to different mechanical pressure as shown by the small arrows.

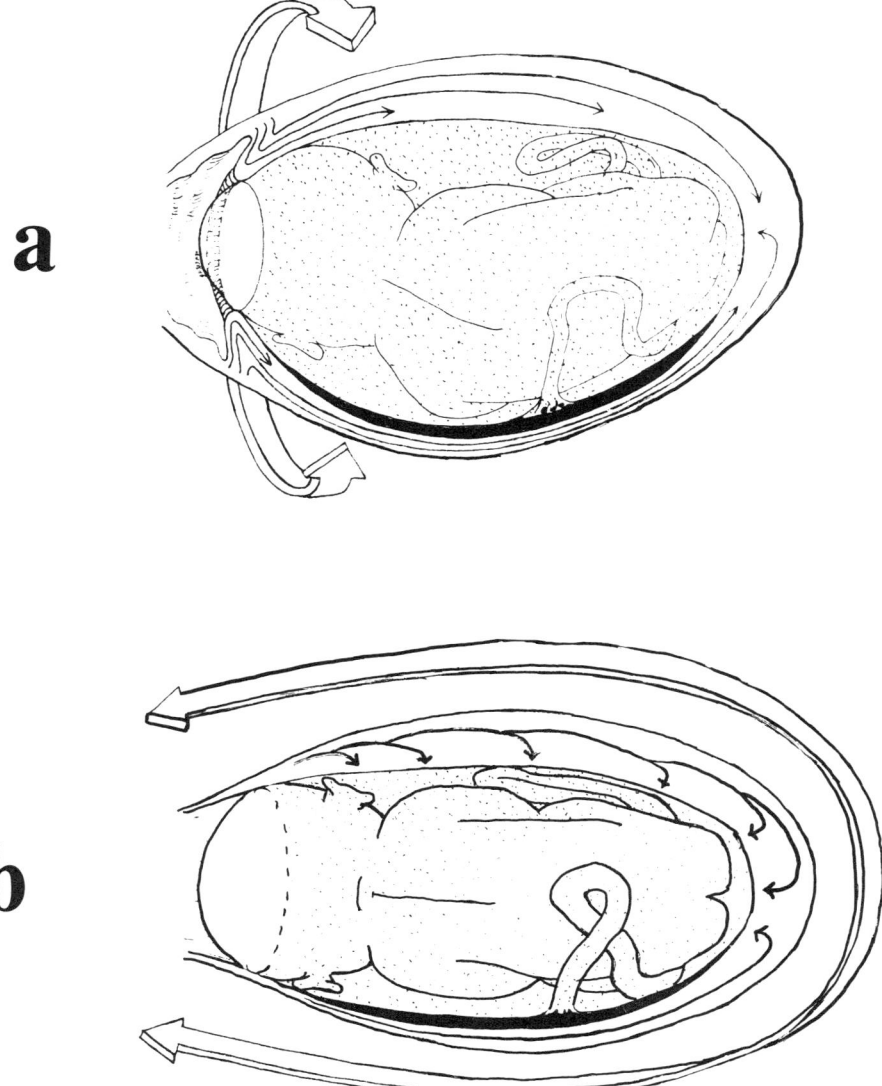

Fig. 5.7. Direction of uterine pressure when the cervix is partially (a) and fully (b) dilated. This is the direction of mechanical force exerted when the uterus intermittently contracts with the added force of muscular shrinkage from involuting muscle bundles. The muscular wall thickens over the extraplacental surface of the egg while it remains thin and distended underneath the placental surface.

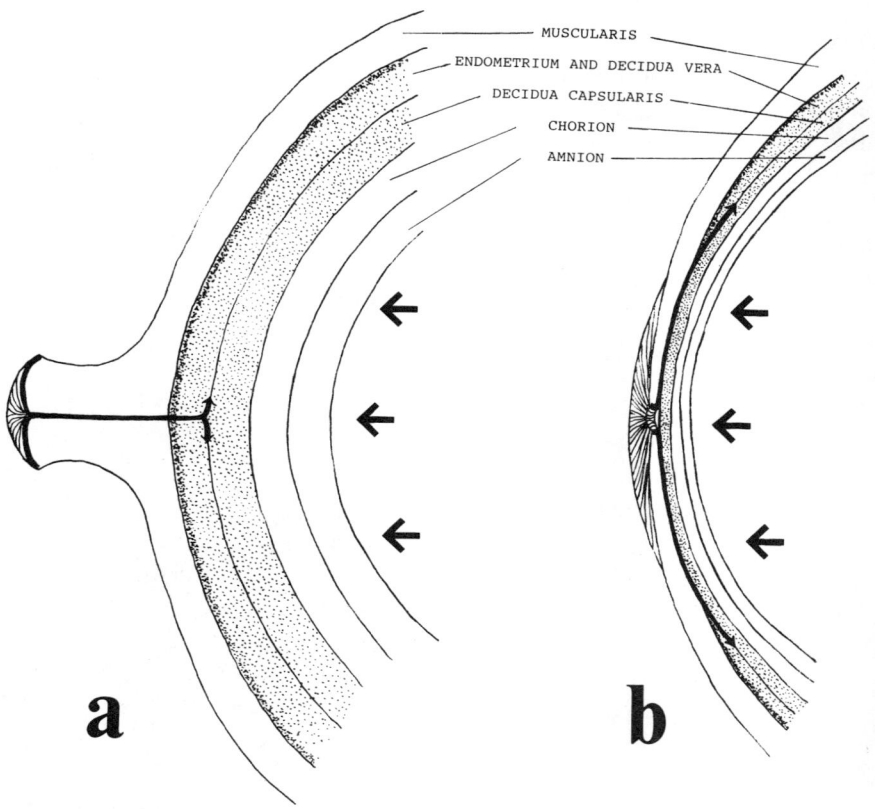

Fig. 5.8. Diagram representing a sagittal section through the cervix and uterus before labor (a) and as the cervix effaces (b). The decidua capsularis at its point covering the internal os is pulled into a widening circle of attachment that stretches it and the amniochorion it contains beyond their limits of elasticity. The three membranes either rupture or balloon through the dilating cervical canal.

contents forming a mucous plug is dislodged into the vagina. As the point of the egg membranes that previously covered the internal cervical os is expanded into the widening circle, the underlying trilayered membrane attached at the rim is stretched beyond the limits of its elasticity and ruptures, a mechanical feat difficult for human hands to accomplish without the use of a sharp instrument.

Morphology of the fetal mass plays a role here. If the fetus is small in proportion to the egg volume, lies transverse, has its shoulder, rather than head, or breech presenting, then the unbraced cervix stretches the decidua capsularis over the dilating internal os to form a tight, watch crystal-like membrane. Amniotic fluid but no fetal part lies against the cervix and the tightly stretched egg membranes. These conditions are precarious. A sudden rupture of the membranes with the resulting gush of amniotic fluid can wash the cord in front of the fetus, a condition fatal to the fetus unless its delivery is immediate. The presence of a fetal head or breech over the internal os, like a stopper floating inside a bottle of liquid obstructs the rapid flow or gush of fluid. The fetal head is a better mechanical block and dilator than the breech. The cephalic presentation of the fetus provides normal labor mechanics.

Whether the trilayered membrane ruptures and releases amniotic fluid or whether it stretches and balloons in an hourglass shape through the cervical canal without release of fluid, the mechanical results in the uterine musculature and the egg are the same. The incompressibility of the egg is lost over its anterior surface, which becomes replaced mechanically by fetal surface. Volume of the uterine cavity is reduced sufficiently to bring pressure from the anterior inner surface of contracting uterine musculature (the surface added by cervical effacement) against the presenting part of the fetus, heretofore completely protected from such force. The reduced volume of the muscular encapsulated egg, by laws of geometry, also reduces its surface and the inner surface of the contracting uterus. With the cervix adding inner surface to the uterine cavity anteriorly and the volume of the uterine cavity reduced, the excess cavity surface is absorbed by the thickening extraplacental uterine musculature posteriorly (see fig. 5.5). This results in a shift of circular-contracting uterine musculature, formally anterior to the fetus, to a position more posterior to the fetus. (Compare fig. 5.5 with figs. 4.2 and 5.6.) The opened egg becomes compartmentalized when the fetal head or presenting part plugs the outlet and seals off escaping hindwater. The seal continues its function of separating the escaping forewater from preserved hindwater, which keeps pressure of fetal small parts from compressing the umbilical cord and placenta during a contraction, maintaining respiratory conditions represented in figure 4.2. The pressure of retained incompressible hindwater against the contracting uterine musculature and fetal mass also maintains the placenta mechanically stretched and distended in its functional state as labor progresses and forward pressure becomes applied to the fetal mass (see fig. 5.6).

The addition of inner surface area anteriorly and its removal in the fundic portion of the uterine cavity do not move the fetus and other egg contents within the uterine cavity. Rather, the uterine musculature and cavity in compartmentalizing and segmenting the egg shift their anatomical relationships relative to the fetus. Some areas, or regions, of the uterus shift their positions while others, such as the area of placental attachment, remain unchanged. The volume of fetus, cord, placenta, and hindwater does not change appreciably after labor is underway. The cervix is pulled over the fetal head or breech into complete dilation when its lips, or rim, completely disappear, converting the anterior circular binding contractions of the uterus into expulsive posterior contractions (see fig. 5.7). The direction of mechanical force in contracting uterine structure is shifted by fetal morphology from a binding pressure surrounding the fetus to a thrusting or pushing pressure from behind, against the hindwater without moving or shifting the fetus within the egg and uterus. Uterine tissue in front of the fetus thins, stretches, and dilates over the fetal wedge, maintaining the fixed fetal position in the egg against the counteracting posterior push as uterine tissue behind the fetus thickens and drives.

The most peripheral layer of uterine musculature, the subserous layer outside the circular binding layer, fixes the uterus to the bony pelvis. The pubovesical fascia attaches this layer of uterine musculature to the symphysis pubis and inferior pubic rami anteriorly, the uterosacral ligaments attach the layer to the sacrum posteriorly, and the round ligaments attach the fundus laterally to the superior pubic rami (figs. 5.3 and 5.4). Contraction of this peripheral muscle layer of the uterine wall in advanced stages of labor, as the cervix approaches full dilation, pulls the uterus and its egg into the pelvis during intermittent contraction. When the cervix becomes fully dilated all uterine musculature becomes positioned and structured to drive the movable egg contents (fetus, cord, and fluid) outside the egg membrane enclosures with the next intermittent contraction. These are the mechanics of hatching and delivery. When the caput becomes visible at the introitus with each contraction, it is not the fetus descending within the soft tissues of the birth canal; it is the entire uterus containing the opened egg that descends within the pelvis as the upper vagina shortens with its dilation. The dilating and shortening of the upper vagina is from outward movement of the forward driven fetal parts within the uterus. The fetus remains stationary within the uterus as dilating and disappearing cervix becomes replaced by dilating and therefore shortening upper vaginal canal. The vaginal fornices are pulled over the stationary presenting part. This intermittent descent and rising of the uterine fundus in the pelvis can begin to occur before the cervix and attached upper vagina are completely pulled, or retracted, into full dilation. Involution at this stage is beginning to spread into the peripheral layer of the uterine musculature attached to the bony pelvis. Upon reaching full dilation, unobstructed delivery, expulsion, or hatching the egg is sudden and brief, accomplished during the span of one intermittent contraction. Labor before delivery can be measured in hours and minutes, but delivery and birth, which is the actual egg hatching in placental mammals, can only be measured in moments.

VI
Function of the Human Uterus during Birth of the Fetus

Delivery and birth of the fetus occur simultaneously under ideal circumstances, but they are not identical. A dead fetus may be delivered, but it cannot be birthed or born. Birth is a physiologic performance occurring solely within the fetus in which respiration is shifted from placenta to fetal lungs instantly without a period of transition or delay. Mechanical delivery, which in all placental mammals includes the actual opening or hatching of the internally incubated mammalian egg, is strictly a function of the uterus. It is an orderly sequence of mechanical steps that first extrudes the fetus into the atmosphere simultaneous with shut down of placental respiratory blood circulation. The uterus delivers or expels the dead fetus in exactly the same manner, utilizing the same orderly steps and the same mechanical means it uses to deliver the live fetus even though birth does not occur. The delivery steps necessarily serve the order of birth functions that occurs within the living fetus. From this perspective it must be recognized that uterine functions serve egg and fetal functions, that uterine functions are to be looked upon as extensions of fetal and egg functions. In all its function and structure, the uterus is a slave to egg physiology. The live fetus is only a part of the living egg under the conditions represented in figure 4.2. Within the uterus all membranes of the live egg, including the fetus, are structured to function together and in harmony. When the fetus dies *in utero*, the contribution of its circulation to the circulation of amniotic fluid ceases. Extrafetal contribution of amniotic fluid also ceases eventually; volume of the amniotic fluid does not continue to increase. The amnion, considered by some to be one source of the amniotic fluid, is actually a membranous extension of fetal body wall. Amniotic fluid accumulation is a part of normal intrauterine fetal growth. If labor does not occur after the egg dies, absorption of fluid from the egg takes place gradually. When the egg within the uterus is opened, as figure 5.7 represents, the fetal-presenting part blocks the escape of amniotic fluid, whether the fetus is alive or dead. Despite the use of the living fetus as a stopper, or mechanical plug, all egg tissues, including the fetus, continue their functions during labor as they did before labor.

Although hours and minutes of time may separate the conditions represented in figure 4.2 from conditions represented in figure 5.7 (first stage of labor), the real structural and functional difference is a mechanically opened but still sealed and unhatched egg, which figure 5.7 represents. By comparison, the uterus converts conditions of the opened egg of figure 5.7 to the conditions in figure 4.1 (delivery) in only a matter of a moment or two. Hatching the egg, i.e., opening the egg, may require several minutes to an hour or so from the beginning of labor (first stage of labor), but hatching the fetus from the egg is comparatively instantaneous. The abrupt loss of all residual amniotic fluid upon delivery of the live fetus, or infant, is the beginning of irreversible membrane desiccation, which cannot occur as long as the amniotic fluid is sealed within the egg. Dessication is the means by which the membranous extrafetal tissue of the egg disappears after the egg is delivered, leaving the fetus as the only egg tissue that continues to remain alive and moist.

The hatched fetus dries some, too. Its skin loses moisture rather than absorbs it. Its lungs engage the comparatively dry, gaseous atmosphere in respiration, evaporating body moisture in the process. Its urine is discarded like the amniotic fluid and the drying extrafetal egg membranes. It absorbs fluid by imbibition through its alimentary membrane. Its body fluid circulation continues as its drying skin shifts its absorptive function to preserving body fluid and regulating body heat loss. The fetus continues to lose fluid and heat the rest of its life as a living individual and must constantly maintain its hydration by continuing to eat food and drink water. Although the undelivered fetus, considered from the mechanical standpoint, is a "volume of membranes" in a "tank of water," functions of membrane surfaces within its body after delivery and birth replace the functions of the extrafetal egg-membrane surfaces through which it functioned while incubated within the uterus. Inside the uterus the fetus is part of a single membrane just as the placenta and cord are also parts of the membrane. Shift of functions from the one area of the egg membrane to the other constitutes birth and brings normally sudden, irreversible changes in structure and function.

Water metabolism and circulation in the compartments of the fetal body continue after hatching as they did before delivery, but their source and exit from the fetus, now an infant, are different. The skin, kidneys, feces, and lungs now lose body water that was received from and returned to the mother's circulation via the placental membrane. After birth, nutrients are absorbed as digested food through its alimentary canal, and gaseous respiratory waste is discharged through its lung membranes. The intimately fused placental membrane within a mechanically separable uterine membrane *in vivo* is more than adequate as a surface for supplying water, nutrients, and nonrespiratory waste-disposal requirements of the fetus within the uterus. It becomes inadequate, however, to handle the respiratory demand of the enlarging fetal brain. The placenta matures *in vivo*, but it cannot grow and further enlarge its respiratory surface beyond maturity. Even if it could enlarge its surface, its oxygen source from circulating maternal blood would probably not be adequate for respiration of a larger fetal brain. Only when the collapsed but fluid-filled fetal lung is converted to the expanded (atmospheric) gas-filled lung after birth, is required respiratory exchange maintained for the future growth and development that is genetically predetermined in the infant.

The grossly imperceptible endometrial covering of the complete egg and its membranes immediately after delivery represented in figure 4.1 should be disregarded from the standpoint of tissue mechanics. Its critical role is in early formation of placental morphology (see chap. 3). As described above, only its structurally prominent contents, the implanted egg, are to be regarded from the standpoint of uterine mechanics and physiology. The egg at term should be considered a single membrane folded and tucked within itself, rather than several separated membranes. In the mechanics of delivery, the outer enclosing portion of the membrane is released from its stretch imposed by the uterus; it becomes irreversibly unfolded, collapsed, contracted, and compacted into a volumetric tissue mass that can be mechanically delivered. Its reduced surface areas then begin to dry and die. This membrane is diagrammed in figure 6.1 in its intrauterine, unborn condition represented in figure 4.2. Within the uterus this single membrane develops with some of its portions and surfaces folded and distended, others folded and compact. All areas are submerged in fluid under pressure to form a protected "package" structured to be opened and "unpacked" in orderly steps and stages that begin with increasing the membrane stretch at a point covering the internal cervical os. Thereafter, pressure and stretch applied to the egg membranes by the shrinking uterus "unpacks" the remainder of the egg. The steps and stages are accomplished by progressive uterine involution applied to the special

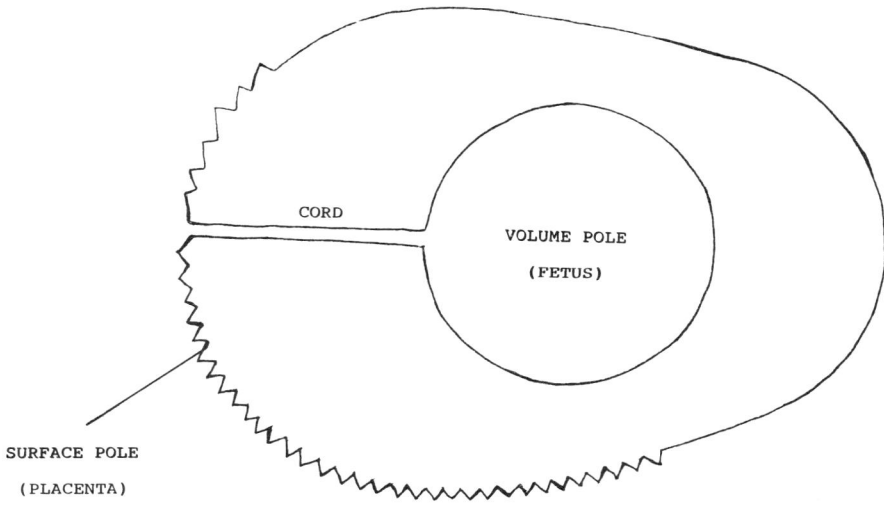

Fig. 6.1. Scheme of the developing egg membranes in mammals. The egg may be considered a single membrane packed into a three-dimensional volume at one pole—the fetus—and expanded into a two-dimensional surface at its placental pole. The poles are united by a vascular connection—the umbilical cord—that grows in a single dimension—length.

morphology of membranous egg structures contained within. The fetus, fluid, and placenta are egg parts morphologically constructed for the mechanics of delivery.

The single-egg membrane is actually a bipolar membrane attached at its surface pole (placenta) and free floating at its volume pole (fetus suspended within fluid trapped within its own skin fold). During development the attaching placental pole structurally and functionally grows and enlarges in two dimensions as tissue surface, while the free and unattached fetal pole enlarges functionally in three dimensions, forming a volume of tissue. The two poles are united by a slender commissure of vascular tissue (cord) that grows or enlarges in only one dimension—height, or length. Birth is the switch of nutrition, water absorption, respiration, and excretion from the attached two-dimensional pole submerged in maternal body fluid (blood circulation of the maternal placenta) to the free volumetric pole drying in the atmosphere. The vascular commissure, the cord, enables this physiological shift in birth to accompany the stressful mechanical steps and stages of egg opening, hatching, and delivery. The morphology of each different area of the membrane has been structured during its development to withstand and yield to mechanical pressure of uterine shrinkage in the sequenced order of labor. Both poles and the commissure of the membrane are all at risk and subject to injury from any mechanical interference with the orderly application of uterine pressure during parturition.

The mechanical stress of delivery is an evolutionary constraint to the orderly birth functions taking place within the egg. Hatching functions, developed long before in the externally incubated egg, are much older functions than the more recently developed mammalian birth mechanics. The gradual change from the older to the more recent involved switch in order of delivery first and hatching later, to hatching immediately preceding delivery. The basis of this evolutionary change was modification of delivery mechanics, the original egg-moving mechanics of the tube.

In both ancient hatching from externally laid, shelled eggs and more recent mammalian birth from an internally incubated egg, lungs of the neonate are dormant and collapsed until the openings of the respiratory tract are exposed to the atmosphere for the initial fetal gasp. Loss of the hard calcium shell rendered the soft egg membranes amenable to rupture or hatching during the stress of delivery in the ovoviviparous stage. The abrupt expulsion of entire unattached eggs that were small gradually changed, as larger eggs became more intimately attached to the endometrium. Delivery of the egg became a lengthier procedure. Maintenance of the more respiratory dependent fetus during delivery of the egg mandated that attachment of the egg be functionally maintained as fetal nares were squeezed through egg membranes to the atmosphere. The fetus and its attachment accordingly became morphologically and mechanically separate parts of the egg membrane joined by loops of cord allowing separate mechanical delivery of the mobile fetus while its fixed, attached membrane surface continued to function. In the human, uterine expulsion of the fetal nares to the atmosphere occurs in the span of a single intermittent uterine contraction.

The structure of the cord, especially its length, endows the fetus with mobility not only for the slight movement, positional shift, and slight change of its shape within the egg, but also for the movement necessary for the fetus to be delivered or moved completely out of the egg and uterus before detachment and delivery of the remainder of the egg, i.e., the placenta and membranes. The cord is lengthy to allow the fetus, or fetal pole, of the egg membrane full mechanical delivery without tension or pull applied to the junction of the cord and the still attached placenta. Separation and delivery of the placenta is also a step in the function of uterine contraction and involution; indeed, it is an even more critical step. It results from reduction of inner surface area of the uterine cavity and mechanical force applied from underneath the placenta by the uterus. Any tension or pull on the placenta from the cord can rupture cord vessels of the fetus. Premature separation of the placenta with devastating hemorrhage can also result. A lengthy cord insures the normal placental separation function in the uterus by preventing any premature pull on the cord as the fetus is expelled.

The intermittent descent of the uterine sling with each contraction can continue for several hours before the cervix finally disappears in complete dilation, but once the cervix is completely dilated, the fetus must be delivered shortly or it will smother. Further involution of uterine musculature beyond full cervical dilation without delivery of the fetus begins to shut down utero-placental circulation, which is the fetal oxygen source. Such loss of uterine circulation and fetal death before its delivery are not from the structural changes alone that occur anteriorly in the cervical region of the uterus, but result from the accompanying changes in structure that occur posteriorly in the fundus. Mechanical pressure on the fetal head, so often accused of being the cause of fetal demise, may be much less of a factor than supposed. After the cervix is pulled into full dilation by thickening posterior muscular wall, further thickening in the posterior wall and fundus compresses uterine vascular channels supplying the placenta. From experience, clinicians have found approximately two hours to be the maximum time the fetus can survive at full cervical dilation. Of course that much time in the second stage of labor only occurs in a complicated labor where fetal delivery is impeded.

The extensive and wide vascular channels of the pregnant uterus are more than are necessary for adequate placental circulation during uncomplicated pregnancy and labor. The circulation of the uterus and placenta is almost always sufficient for carrying nutrients, water, and metabolic waste of the fetus. Brief interruption of these transfers during stress of labor or pressure on the umbilical cord can occur without permanent damage to or sacrifice of the fetus. It is the respiratory requirement of the human fetus at term during labor that places the greatest and most immediate demand on uterine circulation. The extra margin of utero-placental circulation is greatly narrowed during the stressful moments of full cervical dilation in the second stage of labor. The narrowing and compromising of placental circulation and respiration may be the normal means used to bring the newborn fetus to gasp. The delivered fetus, still attached to the placenta by its cord cannot respire or breathe through placental respiration as it did *in utero*; placental circulation in the mechanically collapsed and contracting uterus at this time is too inadequate. Fetal expulsion is scheduled, or timed, in the order of uterine involution to occur at the moment the fetal respiratory system is exposed to atmospheric oxygen (fig. 6.2). This scheduling, or timing, is displayed as order in the progressive structural changes within the uterus, which is determined by fetal morphology that is fixed and changes little; fetal morphology thus complements the structural change within the uterus during labor. When the uterine circulation from muscular compression becomes inadequate to sustain the fetus, the latter is either already delivered by virtue of its own body length (morphology) or it dies.

During this stage of labor (fig. 5.7), the shape of the uterus and the egg change. The axis of the two elongates, and the width narrows somewhat. The placenta is shifted slightly farther posteriorly from the fetus. Even a short cord usually permits this slight change in relative positions of fetus and placenta within the egg enclosure. The uterus and egg tend to assume a cylindrical shape for a moment; then with additional intra-abdominal pressure the fetus is expelled from the egg and from the uterus (fig. 6.2). The extraplacental surface of the uterine cavity becomes greatly

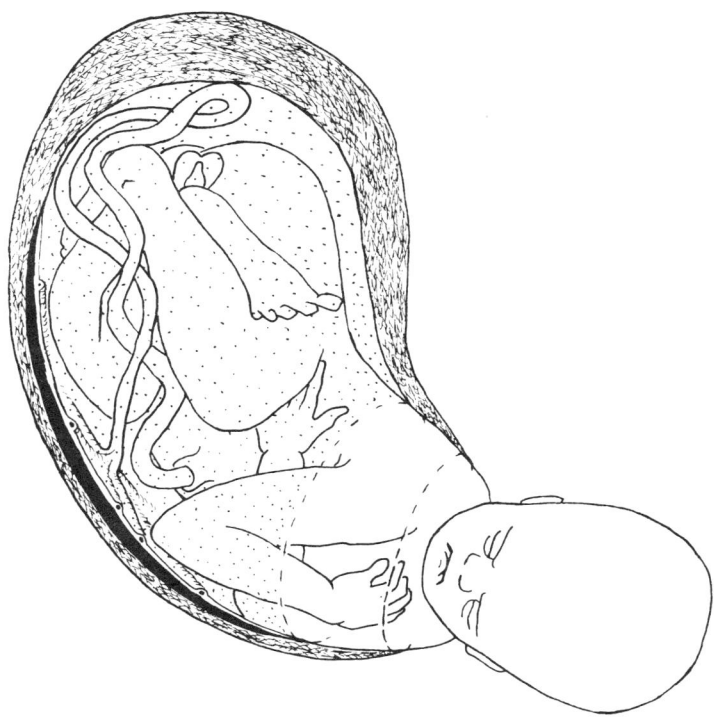

Fig. 6.2. The contracting and involuting muscle bundles thickening the uterine wall in its extraplacental regions. As labor and involution progress, the muscular thickening begins to encroach upon the subplacental muscle layer when the cervix reaches full dilation. The first single intermittent contraction that begins to thicken the subplacental muscular region and close down the placental circulation simultaneously drives the fetus to the atmospheric source of oxygen.

reduced as the musculature thickens and placental blood supply is shut down. It is the large venous channels in the extraplacental musculature that become narrowed just prior to fetal delivery. With the uterus descended into the pelvis and the fetal-presenting part against the perineum, the expulsion of the fetus from the egg and uterus also become its expulsion from the body of the mother. The lower vagina and vulva dilate suddenly and at once as the fetal mass is extruded. The compartmentalized egg membrane is thus unpackaged and its first segment delivered, i.e., the freely mobile fetal pole of the egg membrane.

Fetal and egg structure in relation to uterine structure at this phase of labor is such that a single contraction of uterine musculature delivers the fetal lungs to the atmosphere at the same instant it shuts down uteroplacental circulation. These two mechanical accomplishments, constituting human egg hatching, become a result of a single uterine contraction. There is no circular binding force, separate from the longitudinal, such as that which occurs at the beginning of labor. The entire muscular force of uterine wall contraction becomes positioned behind the fetal-presenting part at this point in uterine involution. With respiration shifted in an instant from placenta to infant lung, the infant gasps as the atmosphere insufflates its lungs. Birth of the infant has occurred, but the uterus did not birth the infant. The uterus only performed the mechanics of opening the egg and extruding the fetal pole of the egg (hatching), mechanics which were under the direction, control, and command of egg morphology and physiology.

This point in the course of labor represented in figure 6.2 represents the earliest point at which the fetus can safely gasp air. As this point is approached (as the cervix approaches full dilation) pressure exerted from behind the fetus by intermittent contractions in the muscular sling, which the uterus begins to form, causes the fetal head to bulge the perineum (fig. 5.3[b]). Residual but diminishing cervical resistance retains the fetus within the egg and uterus. Between intermittent contractions the head recedes only to bulge the perineum again with the next contraction. This can continue an hour or more as cervical resistance begins to disappear. The diminishing utero-placental circulation (it certainly isn't increasing) must continue to be adequate for fetal respiratory needs.

When the labial folds do not close in front of the fetal head between contractions but begin to slide around the fetal crown, the cervix has become fully dilated, offering little or no mechanical resistance. An intermittent contraction drives the fetus, not the egg, from the uterus. This is an irreversible, intermittent contraction of the uterus that does not and cannot stop or be arrested any more than any other intermittent contraction can be stopped or arrested once it starts. There is no difference between this intermittent contraction and all others, even though this particular contraction becomes the one that expels the fetus from both the uterus and the pelvis. The results of this particular contraction are solely the effects of anatomical changes in uterine musculature and its pelvic attachments that convert the uterus into a muscular sling as uterine tissues involute and shrink. The dilating anterior and posterior lips of the cervix disappear as the fetal presenting part fills and dilates the vagina, thus anatomically shortening it (figures 5.3a, 5.4, 5.5, 5.6, 5.7 and 6.2).

The climax of involution, when viewed from the standpoint of fetal welfare, is that point when involution reduces placental circulation below fetal respiratory requirements. That point may be looked upon as occurring in the height of the intermittent contraction that expels the fetus (fig. 6.2). If fetal expulsion is obstructed, the uterus will still expel it and contract behind it, provided the fetus is small and can be physically accommodated by the distended vagina. If the obstructed fetus is large at term, the uterus will rupture and expel the fetus into the peritoneal cavity as it continues its involution. In either case the fetus smothers from lack of placental oxygen due to drop in placental circulation from the contracted uterine musculature. When the placental circulation drops, the fetus must be in the location that enables it to gasp air.

As the fetus is expelled from the uterus, so is that uterine cavity surface formed by the spread of the completely dilated cervix. The contracting uterus drives out the fetus through and against the pressure of the dilated and expended cervical surface; the fetus slides through the tight and widened cervical band of tissue and along its surface. The contracting musculature behind the fetus that drives it forward also pulls the cervical surface over the exiting fetus. The driving posterior musculature is thus anchored to and continuous with the dilated cervix and upper vagina, which in turn are anchored to the bony pelvis (figs. 5.7 and 6.2). But when the fetus is extruded it no longer provides a dilating fulcrum and wedge, and the contracting uterine musculature of the collapsing uterine cavity collects the expanded cervical surface, gathers it into a volume, and expels it along with the fetus into the vagina. This reconstituted cervical volume, unlike the fetus and its lengthy cord, is attached to and continuous with the contracting uterine musculature and, therefore, does not leave the uterine cavity entirely, but protrudes through the circular binding layer as before (fig. 7.2[b and c]). Immediately after fetal delivery the large edematous anterior and posterior lips of the protruding postpartum cervix can be seen in the vagina emitting the umbilical cord lying between them.

VII
Function of the Human Uterus after Delivery of the Fetus

The extrusion of the fetus through the distal vagina and vulva to immediately become a newborn infant has incited wonderment and amazement in all generations from time immemorial. The "miracle of birth" is a constant source of inspiration for writers and poets, and in some form becomes the center of dogma in religions. The infant's first gasp and cry dispel all tension in nonprofessional attendants, who immediately relax and rejoice in their excitement, completely oblivious to the important role of remaining uterine function in human reproduction. Writers, storytellers, and the clergy as well as the lay public have derived little inspiration from delivery of the afterbirth. When death of the mother results from failure of uterine function in this stage, the death is associated with the fetal birth, and the mother is reported to have died from childbirth or "in childbirth." The professional attendant with experience becomes tense during this period immediately following delivery of the fetus, because he is well acquainted with the consequences of failure in uterine function at this crucial and dangerous point in human parturition. Function of the uterus after delivering the fetus is the most important and critical moment in all the physiology of human reproduction. It prevents catastrophic and fatal hemorrhage in the human mother as the placenta separates. This function of the uterus has brought humanity into existence, and on it depends survival of the present human race and its descendants. It is a mechanical performance and is the high point, or climax, in separation and delivery of the intimately attached human egg.

Extrusion of a living fetus can and does often fail to occur. A live birth at every parturition is not necessary for human survival. On the other hand, delivery of all dead fetuses by parturition functions of the uterus is essential and fundamental to successful human reproduction and survival. If a dead egg is not removed by delivery but is retained, reproduction of that mother is halted. The human reproductive cycle, a reproductive cycle in which only one fetus, or egg, is normally hatched, is polyestrous, biologically more fundamental than the monestrous cycle. A cycle resulting in delivery of a fetus, dead or alive, is followed by another cycle with overwhelming odds of eventual live births resulting. The opposite is the case when function of the uterus following fetal delivery is considered; the placenta in all cases of both living and dead fetuses is functionally a dead organ at delivery. From the biologic standpoint delivery of the placenta must never fail. Sacrifice of the mother during normal parturition resulting from postpartum hemorrhage, infection, or both would be more than overwhelming odds against human survival. Indeed, it would be incompatible with human origin and is an evolutionary impossibility. Frequent sacrifice or loss of the fetus is necessary and evolved as a necessary component of human reproduction, but such frequency in loss of the mother from failure of placenta-delivering physiology of the uterus is not biologically and physiologically tolerable. The two most visible uterine functions of human egg hatching and delivery are delivery of the fetus and delivery of the placenta. As difficult as it may be to realize, the delivery of the fetus is a relatively incidental event in the course of labor and is the lesser of the two functions in importance to human survival. It has been said that we have been placental mammals for 130 million years (Jolly, 1972). It does not appear from the human species that placental mammals are on the way out by being replaced by another form of mammal. The evolution of the human placenta accounts for the evolution of the human fetus, but evolution of the human placenta in turn is also an evolutionary result, the result of evolving physiology that provides for its safe, mechanical separation from the uterus. Preserving the life of the mother is the chief function of the uterus during the dangerous ordeal of the third labor stage in human reproduction. Evolution of this safe, mechanical separation evolved as the principal component of the fundamental physiology of the human uterus, the human reproductive cycle, and the reproductive cycle in upper primates that menstruate. It is a more fundamental part of the marvelous and amazing miracle of the human birth process than the visible expulsion of a living fetus.

The mechanical extrusion or expulsion of the fetal mass is not much more mysterious or difficult to comprehend than the mechanical movement of food or fecal mass by contraction of smooth muscle in the bowel. The fetus with its placental oxygen supply cut off and thirsting for oxygen gasps air on its own if its respiratory system is in contact with the atmosphere, i.e., the fetus contains its own breathing physiology (see birthing physiology in chap. 6). What is mysterious and marvelous in uterine function, however, is the safe delivery of the intimately attached human egg after it has been emptied or hatched when guarantee of the mother's safety becomes paramount.

The sequence of extruding the fetus first, permitted by a lengthy cord, immediately dismisses the fetal respiratory burden as a physiologic factor in delivery of the placenta.

What remains to be delivered after the fetus is the attached egg pole and all of the trilayered egg coverings, which are the amnion and chorion contained within the compact layer of endometrium (fig. 4.1). So intimately fused are the egg membranes with the endometrial layer in the human that separation and delivery of the entire surface endometrium is accomplished. This is in contrast with the case in lower-ranking mammals, where cleavage between egg membrane and endometrium takes place instead. The enormous blood supply to the separated endometrial layer, which contains and previously nourished the human egg at term through minute arteriolar capillaries, is staunched by arteriolar sphincters in the contracting and shrinking inner layer of uterine musculature. How continuing contraction and involution of the uterus, following delivery of the fetus, separates and delivers the remainder of the egg—the placenta and membranes—and simultaneously staunches its blood supply underlies the fundamental reproductive physiology of the human uterus. It is the same tetanic uterine contraction, squeeze, and cramp that produces dysmenorrhea and the endometrial ischemia phase of the menstrual cycle. The involution of this contraction does not begin after delivery of the infant at term; it is already in progress when the uterine cavity collapses during delivery of the fetus. It is simply the continued involution of uterine musculature that extruded the fetus, which in turn was a continuation of involution that opened the egg.

These stages and functions of uterine involution are imperceptibly brief and transient in the absence of the implanted egg during the menstrual cycle. Labor after egg implantation is one continuous episode of involution in uterine musculature that begins in one region and, diverted by egg morphology, gradually spreads into other regions of the uterine musculature to finally include the area beneath the attached placenta. In the previous chapter its order was shown to complement the structure of the egg, i.e., fetal and placental morphology, in an order designed to open, expel, separate, and expel contents of the uterine cavity. The first sequences, or phases, of uterine involution that opened the egg and delivered the fetus have already been presented.

The last and final area of the uterus included in involution is the muscular layer lying beneath the attached placenta. It begins to involute as the fetus begins to extrude, since its uninvoluted state is absolutely essential for fetal respiration during fetal expulsion. Contraction and involution of this layer separates and delivers the placenta but does not stop. Contraction and involution of the entire uterus also separates the rest of the chorion and egg membranes when it detaches the whole superficial layer of endometrium. Still it does not stop involuting and contracting. Labor continues hours, days, and weeks after delivery of the egg membranes and placenta. It ends when the uterus is small enough to again mechanically manipulate sperm and the tiny egg to establish a new pregnancy. The next pregnancy can only occur three weeks or later after delivery of the placenta, but in some instances has been known to occur as early as two weeks. Labor, or involution of uterine musculature, starts after continued endocrine growth stimulus to uterine musculature is withdrawn, and once started, labor is difficult to stop or arrest.

Arresting or stopping labor are terms that usually refer to temporarily delaying progress of labor. These terms are used in efforts to delay the advance of labor in cases of premature labor, hoping to allow the fetus more time for further growth and development. Once labor is in progress with uterine contractile elements involuting, shrinking, contracting, and constricting uterine blood channels, further vascular enlargement in the uterus from growth of the egg and fetus is unlikely. Even if artificial endocrine growth stimulus is successful in instituting growth in the involuting musculature, which would stop involution, or labor, this does not guarantee a resumption of endocrine supply from and growth within the egg (placenta and fetus). Whatever caused the egg to withdraw its endocrine growth stimulus to the uterus and bring on labor must first be corrected in order to arrest or stop true labor. Probably the only truly arrested labor is a missed labor. Cases of women in their eighties carrying a calcified fetus in their womb are well documented. In these cases progress of the pregnancy was arrested; i.e., growth of uterine musculature ceased, probably gradually, when the placenta died and its estrogen and progesterone levels slowly rather than abruptly dropped. Perhaps the drop in estrogen and progestin levels was too slow and gradual to bring on involution of labor and egg delivery. Intermittent contractions of the uterus eventually ceased, and shrinkage of musculature occurred slowly and gradually over a comparatively lengthy period of time. The quiescent uterine wall simply shrunk and thinned about its dead, encapsulated contents. The organ became one of absorption and calcium deposition; mechanical action of its delivery became deferred and remained potential.

With the fetus and its welfare being secondary to more important phases or stages of involution that follow fetal delivery, uterine involution next reaches that point when involution and contraction of the uterus completely clamps off the endometrial blood supply to the separating endometrial layer. This is the blood supply to the superficial endometrial layer that contains the entire egg even while it is attached within the uterus. These two events—(1) uterine contraction, i.e., contraction of the entire organ including the subendometrial muscle layer, and (2) endometrial ischemia—like the two previous events of reduction in placental respiration and fetal delivery—are not just simultaneous events or functions but are actually two accomplishments by the same intermittent uterine contraction. Progressive change in morphology of both egg and uterus brings about these different accomplishments by a single, mechanical contraction. One single intermittent uterine contraction spreads to and includes contraction of the subendometrial muscular layer operating the spiral arteriole sphincters that control arterial blood circulation to capillaries in the superficial endometrium and also mechanically separates and extrudes endometrial surface containing the placenta. The placenta is thus separated and delivered by the endometrial blanching or "staunching" contraction (fig. 7.1).

With the delivered infant crying and its cord still attached to the placenta inside the uterus, the extruded portion of the cord begins to lengthen as the uterine walls thicken and the cavity begins to shorten (fig. 7.2). The continued involution of uterine musculature continues to reduce the volume of the uterine cavity by the same means it reduced cavity volume in delivering the fetus, by diminishing its inside surface area. Only now the thickening uterine muscular walls constrict and flatten uterine blood channels—the uterine plexus of veins—which yield additional volume and space within the uterine wall for accommodating the thickening muscle bellies. This absorption of volume and space by the thickening uterine wall permits greater control and direction of surface area reduction, and thus volume reduction, inside the uterine cavity.

In its distended state at term during labor the uterus is ovoid in shape. When collapsed and empty after fetal delivery the smooth muscle sheets forming the muscular walls cannot contract symmetrically on all sides of

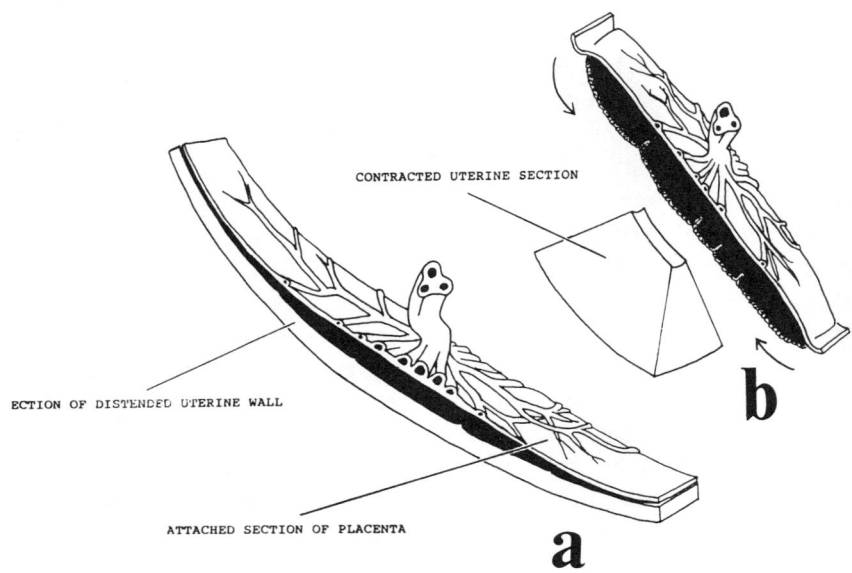

Fig. 7.1. Diagram of a segment of uterine wall with corresponding stretched and functioning placental segment attached (a) and contracted (b). When the subplacental muscular wall contracts after fetal delivery, it mechanically reduces the area of placental attachment over the inner surface of the uterine cavity. Contraction of the corresponding placental segment, because of villous growth, becomes locked into a volume with greater surface. The net shearing effect of the mechanical force separates the two surfaces.

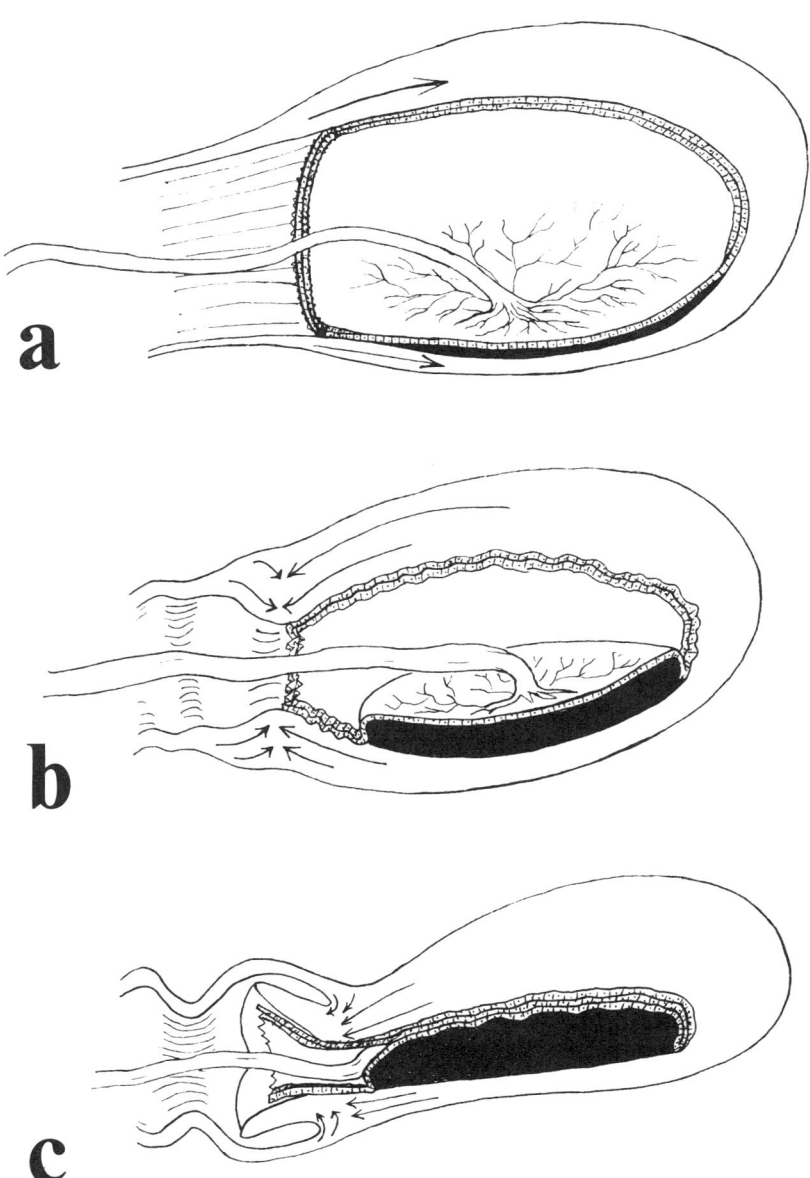

Fig. 7.2. Diagram of extruded cord after fetal delivery. During expulsion of the fetus (a) the cervical lining of the uterine cavity follows the fetus and is delivered along with the fetus into the vagina as the reconstructed cervix (b) and (c).

the cavity to reduce its surface and volume. Just as the walls of a symmetrically contracting globular or spherical muscular chamber would have to contract to a center point with zero surface to acquire zero volume (fig. 7.3[a and b]), the ovoid pregnant uterus would have to contract to a center line without surface. Such contraction of a muscular chamber through thickening of its muscular walls is not mechanically or physically and geometrically possible in an extruding squeeze. The thickness and rigidity of the tightly contracted walls would leave some form of hollow cavity. To obliterate such a cavity, gathering rugae of the inner lining tissue actually project inward and replace cavity space. This is seen in other body cavities such as the stomach and bladder.

For complete mechanical emptying, the uterine cavity volume must be reduced to zero, yet for the next pregnancy residual endometrium attached to the muscle wall, the endometrium basalis, must be preserved and present. To produce a completely empty cavity of zero volume with a preserved cavity surface and its blood supply, the contracting uterus, in the third stage of labor, contracts into a flattened cavity with its two surfaces tightly opposed. Most contraction takes place in the anterior and posterior walls of the uterus rather than at its hilus or vascular margin (fig. 7.3[c]).

The lengthening cord protruding through the introitus after fetal delivery is followed by a lower abdominal bulge as the contracting uterus rises out of the pelvis. The contracting but collapsed muscular sling, in reducing its inner surface first, assumes a more globoid shape from additional contraction in the circular binding layers. The contraction gathers the formerly stretched and distended placental surface within the superficial endometrial layer into a volume of tissue, the contracted placenta, which the flattening uterine cavity briefly accommodates (fig. 7.2). The morphological change in contraction of the stretched placental membrane to form the placental organ is actually a description of the contracting and separating endometrial surface. As human pregnancy is totally intramembranous, contents of this endometrial surface, when the fetus and fluid are released, account for its gross appearance as a placenta.

Conversion to a volume by the collapsed and contracted membranous placenta through uterine contraction and involution is a result of its structure, morphology, and development. The placenta develops and functions as a stretched membrane under mechanical tension. At the end of the fourth month, when the uterine cavity becomes obliterated mechanically by the fusion of the endometrium capsularis and endometrium vera, the intramural egg in its spherical contour has developed its villi and endometrial sinuses in a radial direction. In a general way the villi and sinuses point toward the center of the uterine cavity and are perpendicular to the endometrial surface. But as pregnancy growth advances under pressure of uterine musculature, the radial structures consisting of capillaries and glands become flattened as the egg acquires additional surface over its placental area. Secondary villi advance into the flattening venous sinuses in an everting manner (figs. 7.4 and 7.5). The addition of tissue surface in the form of secondary branching villi in the expanding decidua compacta provides irreversible contractility in the placental membrane. When the placental surface is gathered in the diminishing cavity surface by uterine contraction in the third labor stage, the placenta begins to contract also. The villi tend to invert as the placenta compacts, bulging the fetal surface of the organ into the accommodating cavity space. But as the placental membrane surface contracts, it is blocked from further contraction of its stretched elastic state by the proliferated villi, which lock into a compacted and incompressible volumetric mass with a fixed maternal surface area (fig. 7.1). The attached uterine surface, however, continues to reduce its area, but the contracted and compacted placental tissue cannot further invert to reduce its surface in accommodation. This shearing mechanical action applied to the endometrium compacta now in an irreversibly expanded state causes it to split in the capillary level above the endometrium basalis (fig. 4.5).

The combination of fixed placental surface and marked reduction in area of placental attachment within the endometrium separates the superficial endometrial layer containing the placenta. The same shrinkage and contraction of the inner uterine surface also eventually separates the extraplacental endometrial surface that contains the fused decidua capsularis and vera. The intermittent uterine contraction that separates the placenta is the contraction that occurs in the muscle layer in the uterine wall lying beneath the placenta. It is the same contraction that clamps off the arterial blood supply to placental sinuses—the spiral arterioles (fig. 7.6). The delivery of the placental volume, like delivery of the fetal volume, becomes the mechanical function and accomplishment of a single intermittent uterine contraction.

Postpartum hemorrhage results when the inner muscular layer, the muscle layer adjacent to the endometrium, relaxes. When the portion of this layer lying beneath the placenta contracts and separates the placenta, the remaining portion of the layer, approximately two-thirds of inner surface of the uterus before labor at term, remains relaxed providing cavity

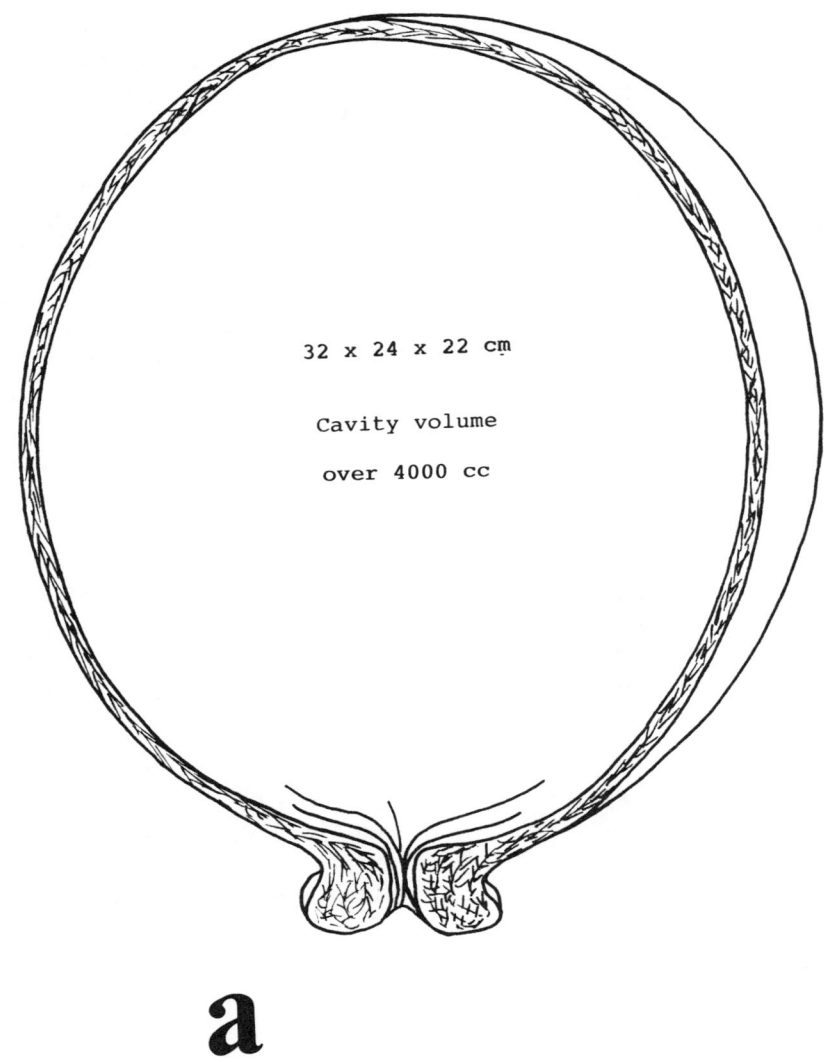

32 x 24 x 22 cm

Cavity volume

over 4000 cc

a

Fig. 7.3. Mechanical forces exerted by the muscular wall of the uterus in separation and delivery of the placenta. The geometry of uterine mechanical force is determined by uterine morphology. Although the gravid uterus at term is not completely spherical in shape (a), its muscular walls contract, when delivering the placenta, into a muscular sphere. The contraction is not uniform as would be the case in (b) if uterine morphology were spherical, gathering the inner layers into inward-projecting masses. The anterior and posterior walls of the postpartum organ thicken greatest (c), producing a flat cavity with a preserved surface.

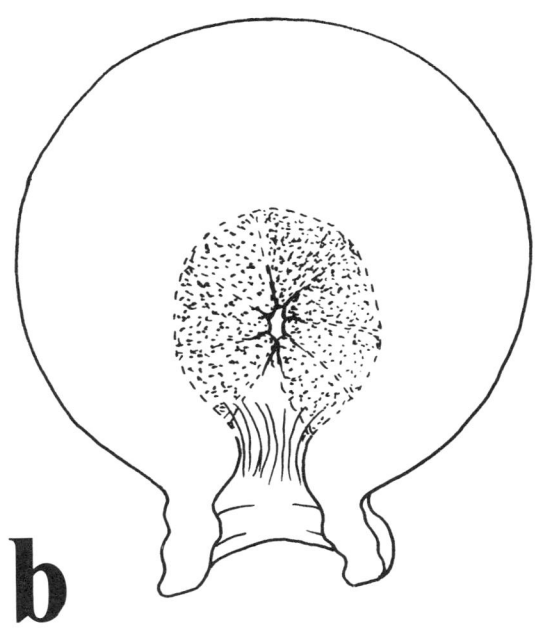

b

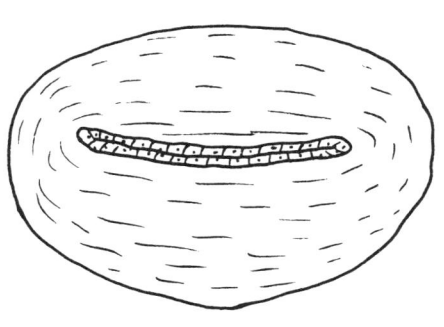

c   19 x 12 x 8 cm

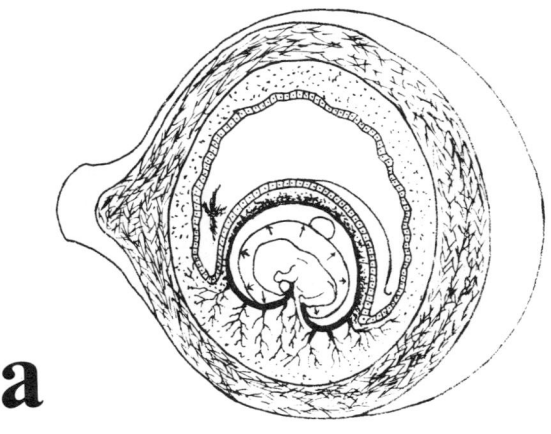

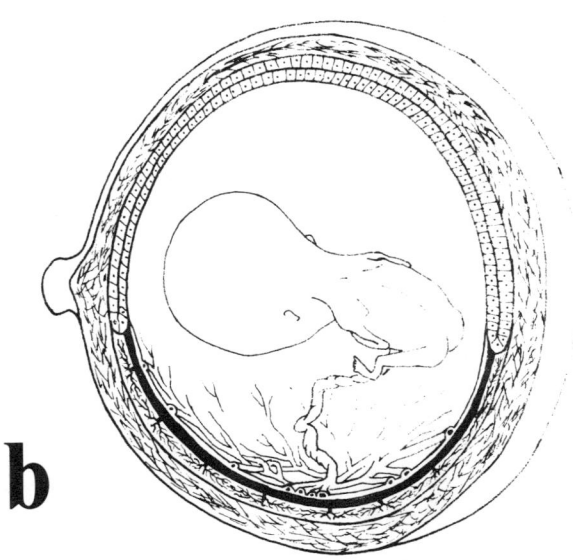

Fig. 7.4. Diagram showing flattening of radial structure. The early placenta develops under tension of the expanding amniotic fluid (a) and acquires additional surface area in the expanding endometrium through flattening of radial structure (b).

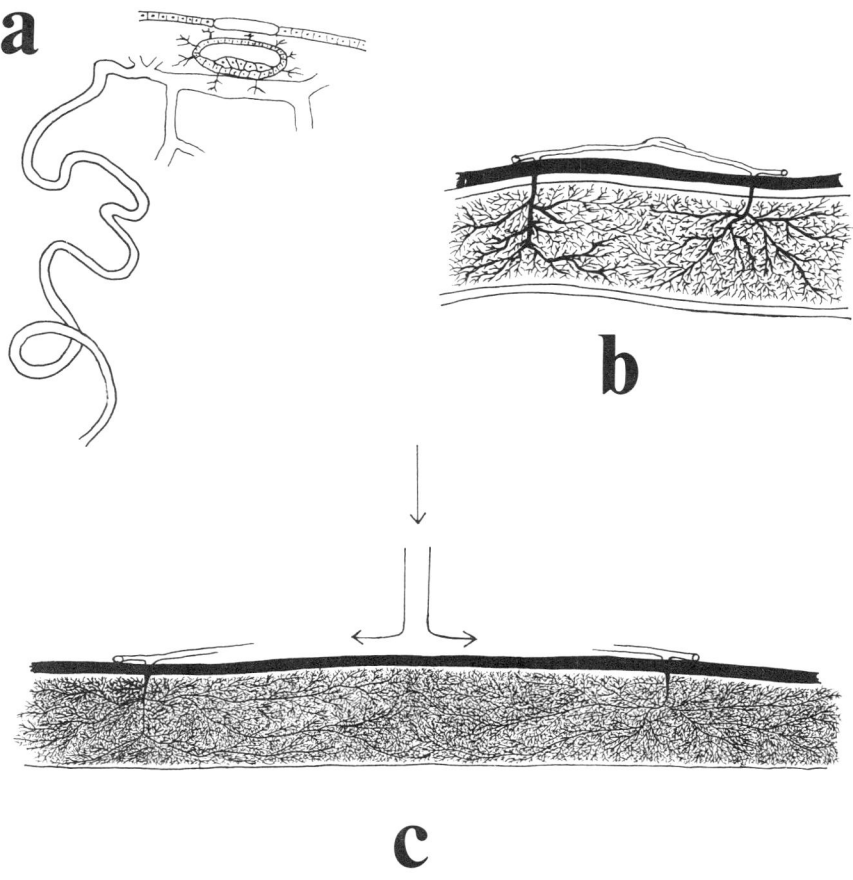

Fig. 7.5. Diagram of advance of secondary villi. The newly embedded human egg immediately begins villus formation (a). After the growing and enlarging egg fills up and seals off the uterine cavity at the end of the fourth month the villi constitute a fully formed placental organ (b). Additional placental surface is then acquired by the expanding organ through secondary branching of the earlier formed villi (c).

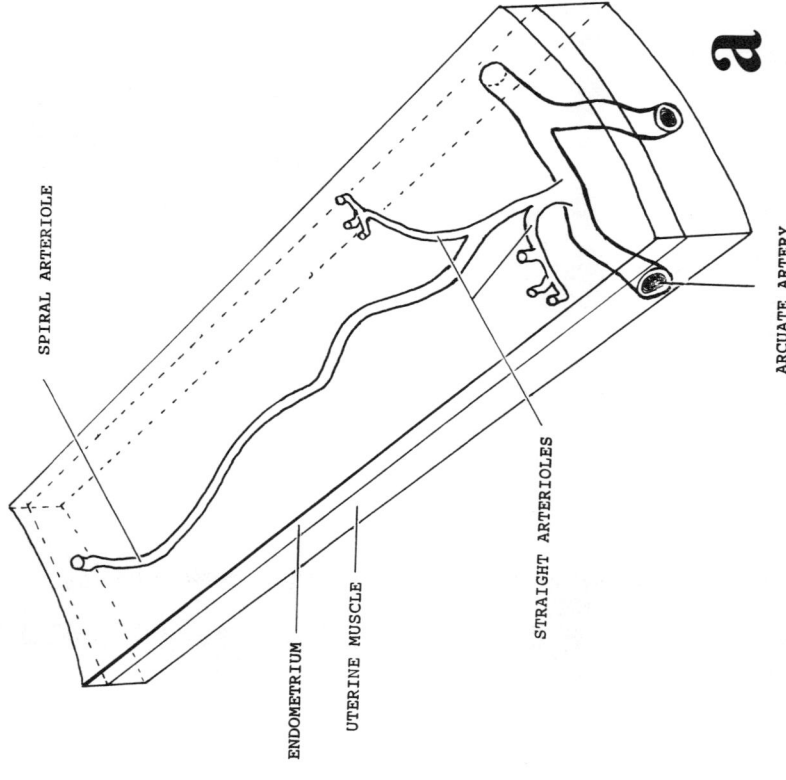

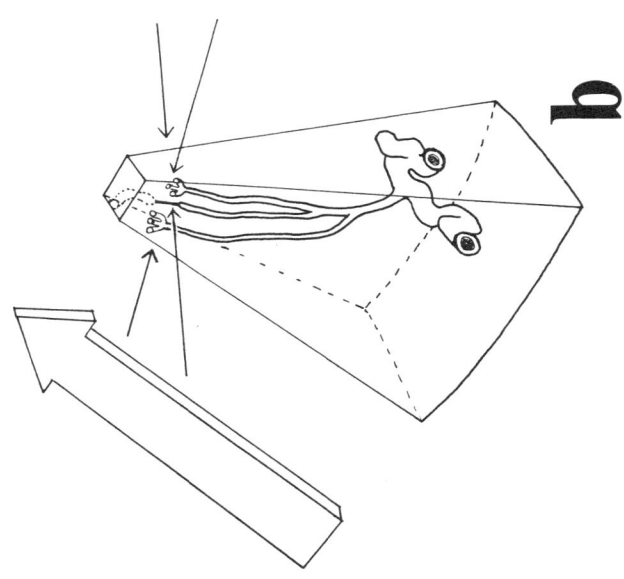

Fig. 7.6. Diagram of a segment of uterine wall and endometrium beneath the placenta (a) and contracted after delivery of the placenta (b). Thickening of the inner muscular wall sphincters the spiral arterioles (small arrows) that previously supplied the surface endometrium and the placenta while simultaneously expelling the placental mass (large arrow).

space that briefly accommodates the gathering and inward projecting placenta. This relaxed portion of the inner-muscle layer, during these brief moments that the uterine cavity accommodates the separated placenta, continues its blood supply to the gathering, compacted, and unseparated extraplacental decidua compacta. This latter is the double layer of the two fused surface layers of the decidua compacta with the inner layer containing the extraplacental amnion and chorion (fig. 4.3). This remaining endometrial surface is supplied by the spiral arterioles, just as the placental area is supplied, but is without villi and placental sinuses. Whether these extraplacental vessels of the endometrium supply the amnion and chorion the means of producing amniotic fluid during pregnancy has not been proved. All blood vessels supplying the endometrium are small, precapillary arterioles and venules that provide the uterus the means of controlling the blood supply to the cleaving capillary bed that joins the maternal placenta with maternal blood circulation.

Although volume of the separated placenta is accommodated temporarily in the flattened uterine cavity as the uterine ovoid rises out of the pelvis, the uterine mass then sinks into the pelvis as the thickening anterior and posterior walls and fundus of the uterus extrude the placenta from its flattening cavity. The thickening and contracting anterior and posterior uterine walls in reducing their surface area also separate the remainder of the extraplacental endometrium compacta containing the remainder of the chorion and amnion. The compacta is contracted into a volumetric mass that separates in the cleavage plane formed by flattened endometrial glands. The muscular thickening reaches its maximum in the fundus and spreads anteriorly as both cavity walls thicken and flatten the uterine cavity to produce zero volume and expel the remaining volume of surface decidua containing the egg membranes. The subplacental muscular layer is pulled from its location on one wall around the separating placenta, which it delivers as it forms the subendometrial muscle layer of the empty uterine cavity. The cavity is thus completely obliterated mechanically with preservation of its endometrial lining and intact endometrial blood supply that can now form a new surface.

VIII
Function of the Human Uterus between Pregnancies

Contraction, involution, and shrinkage of the uterus continue for at least two or three weeks postpartum. Although some clinicians distinguish a fourth stage of labor as the contraction of the uterus during the hour following delivery, intermittent uterine contractions that can be very painful cramps continue for several hours after delivery and gradually diminish in severity. As the infant begins to nurse, the mother can often feel the cramps stimulated by nursing. After a day or two she may feel a weightiness in her pelvis rather than a cramp when the infant nurses. Right after delivery the contracted uterus is more globoid and is the size of an infant's head but rapidly reduces in size during the next week to ten days. After approximately ten days, the uterus approaches its average prepregnancy size. The edematous cervical lips shrink to their nodular, nonpregnant shape protruding into the upper vagina, the endometrial surface is reduced in area and repaired, and the organ resumes its sperm transporting and menstrual functions.

The postpartum uterus, a tumorous mass of tissue ten centimeters in diameter and easily palpable in the lower abdomen right after delivery, reduces so rapidly in size it produces metabolic effects. The involuting musculature returns nitrogen and water to the circulation, as the mass of uterine muscle fibers shorten and thin to their prepregnancy size. As marvelous as this may appear to be, the real marvel in the organ at this time is the repair of the shorn endometrial surface, the membrane surface that must be preserved and repaired for a subsequent pregnancy. Each step and stage of uterine involution appears more marvelous in its accomplishment of physiological tasks, until the organ reaches its return to the delicate mechanical functions of moving microscopic sperm and tiny egg particles over mucous membrane surfaces.

Immediately after delivery of the placenta and membranes, the obliterated uterine cavity theoretically does not permit even the introduction of a finger. A cross section of the uterus should reveal simply a central transverse slit in the muscle mass, rather than a cavity (fig. 7.3[c]). At autopsy, in cases of maternal death after delivery of the placenta, the excised postpartum uterus can be redistended almost to its prelabor size (figs. 4.2 and 7.3[a]). When the redistended organ is opened, one would expect to see very clearly the area over which the attached placenta stretched. It can be a disappointment to find hardly any evidence of the discrete area where the placenta attached. The entire surface of the endometrium is gone. A concentration of only a few large vessels in the muscular wall may be the only hint of the general location where the placenta attached. The reason for this finding is the architecture and function of the endometrium. Its decidual surface is a capillary bed that the implanting egg enters and expands into its enormous villous-sinus respiratory surface. In pregnancy and during the first and early part of the second stage of labor, the functioning placenta is still located inside this stretched and distended surface layer of endometrium (decidua compacta), with its vast arterial blood supply provided solely by the terminals of the uncoiling spiral arterioles. These vessels uncoil during growth and expansion of the surface endometrium in pregnancy to guarantee a continuing blood supply to the capillaries that flatten, widen, and shift in their lateral dimensions to form the placental sinuses (figs. 7.4 and 7.5). The placental surface area and space are thus acquired within the decidua compacta—the endometrial surface layer—not within the uncleavable, deeper endometrium basalis. This entire superficial endometrial layer is missing in the autopsied uterus immediately after the third labor stage. Since only very fine microscopic vessels, i.e., capillaries, supplied the deciduous, endometrial layer in forming the placenta, little evidence of placental location remains. The endometrium basalis remains histologically unchanged throughout pregnancy and labor, although it necessarily undergoes some positional shift. The basalis is attached to the distending and growing muscularis and is supplied by straight arterioles (uncoiled) and venules. The straight arterioles and venules and the endometrium basalis that they supply accompany the distending and growing muscularis, but the widening distance and space between each set of straight vessels become occupied by growth of the superficial endometrium, containing the expanding placental sinuses supplied by the uncoiling spiral arterioles (see fig. 4.5).

Neither the straight vessels nor the spiral, or coiled, vessels enlarge in their diameters during pregnancy, even though the spiral vessels uncoil and transmit the enormous fetal oxygen supply. It is necessary that these small endometrial vessels remain small throughout pregnancy and the entire reproductive cycle not only to facilitate their safe mechanical severance in placental separation, but also to facilitate the mechanical staunching of their blood flow. From the standpoint of nourishing the egg and fetus during pregnancy up through early second labor stage, the endometrial blood vessels could very well have enlarged, but more than egg nourishing physiology has become the evolutional constraint here. Where enlarged endometrial vessels would fail is in the repair and healing of the shorn endometrial surface after the placenta and membranes are delivered. In order for the denuded endometrial surface to reform, it must first heal, and to heal it must reepithelize. Although the new epithelial surface extends from the residual glands of the basalis, the blood supply for the proliferating and reepithelizing stroma of the new compacta is from the healing spiral arterioles. Placental cleavage may be looked upon as occurring along the junction line of the capillaries supplied by the spiral arterioles on the upper side and the straight vessels on the lower side. The blood flow through the spiral arterioles can only be returned intermittently through the newly forming surface capillaries to the capillaries of the endometrium basalis during endometrial healing. Enlarged diameter of the spiral arterioles during pregnancy would not be compatible with capillary proliferation and tissue surface repair. All venous return of the placenta and endometrium is through the small straight venules of the basalis.

A comparison can be made with the use of a tourniquet. The tourniquet must be loosened at intervals with bleeding through the finer vessels at the injury site in order to preserve healing tissue; complete and permanent occlusion of large vessels would completely suffocate and kill the distal tissue. Also, contrariwise, intermittent release of a large vessel sphincter would return blood flow to all the distal tissue at once with hemorrhage through the damaged vessels. The uterus controls endometrial hemorrhage by intermittently releasing and sphincterizing each individual spiral anteriole at different times during the repair process. Thus some areas of the healing endometrium enjoy a period of returned blood supply while other areas are kept ischemic. Then ischemic areas are blushed with their circulation while previously blushed areas are temporarily rendered ischemic. By this means the uterus controls endometrial bleeding during its postpartum healing period. During healing the endometrial surface must not be suffocated, killed, and shed, as it is during the endometrial ischemia phase of the menstrual cycle and during the third labor stage, but must be preserved, nurtured, and stimulated to grow and heal. The fundamental physiology of the human uterus, however, is to blanch, suffocate, kill, and shed its endometrial surface. All healing of surface endometrium in the human uterus can only be done through this physiological sequence of blanching, suffocating, and shedding it intermittently. Its blood supply must continue throughout and after healing. Sphincter action of the muscularis around the base of each spiral arteriole intermittently relaxes and tightens in order to provide sufficient blood supply for the healing and regenerating endometrial surface without extensive hemorrhage from the shorn membrane surface. Endometrial capillaries continue to regenerate with stroma proliferation until the endometrial surface is completely repaired. Blood escaping from the damaged unshorn vessels during menstruation may assist in separating the adhering and autolysing capillaries from the repairing surface underneath. This same repair takes place in the postpartum organ (see chap. 3). Blood escaping from the damaged unshorn vessels in the menstrual cycle may assist in separating the adherent membrane (see chap. 3).

Figure 8.1 illustrates the different functions in spiral and straight vessels. Contraction of the entire postpartum uterus is toward flattening and obliterating the uterine cavity space from the fundus forward. Contraction of the subendometrial muscle layer constricts the sphincters at the base of the spiral arterioles, leaving the straight vessels, both arterioles and venules, relaxed and open. This is the fundamental biologic mechanical action for which the human uterus is constructed and is available to be exercised in part or wholly in all stages and phases of its performance in reproduction. It accounts for the endometrial surface ischemia that occurs just before menstruation, as well as the staunching and control of hemorrhage at separation of the placenta and the intermittent blush and blanch of the healing postpartum endometrial surface. For the inner subendometrial layer of the uterine muscularis to relax, all other uterine muscle layers must also relax and be partially distended as they are in pregnancy. When this occurs just after removal of the placenta it results in postpartum hemorrhage. The hemorrhage is entirely from the spiral arterioles due to relaxation of the sphincter function in the muscularis at their bases. As long as the entire uterus remains contracted, the sphincter action in the innermost contracted subendometrial layer of uterine musculature prevents extensive postpartum hemorrhage by producing ischemia of the endometrial surface. Contraction of the uterus immediately after placental delivery is a

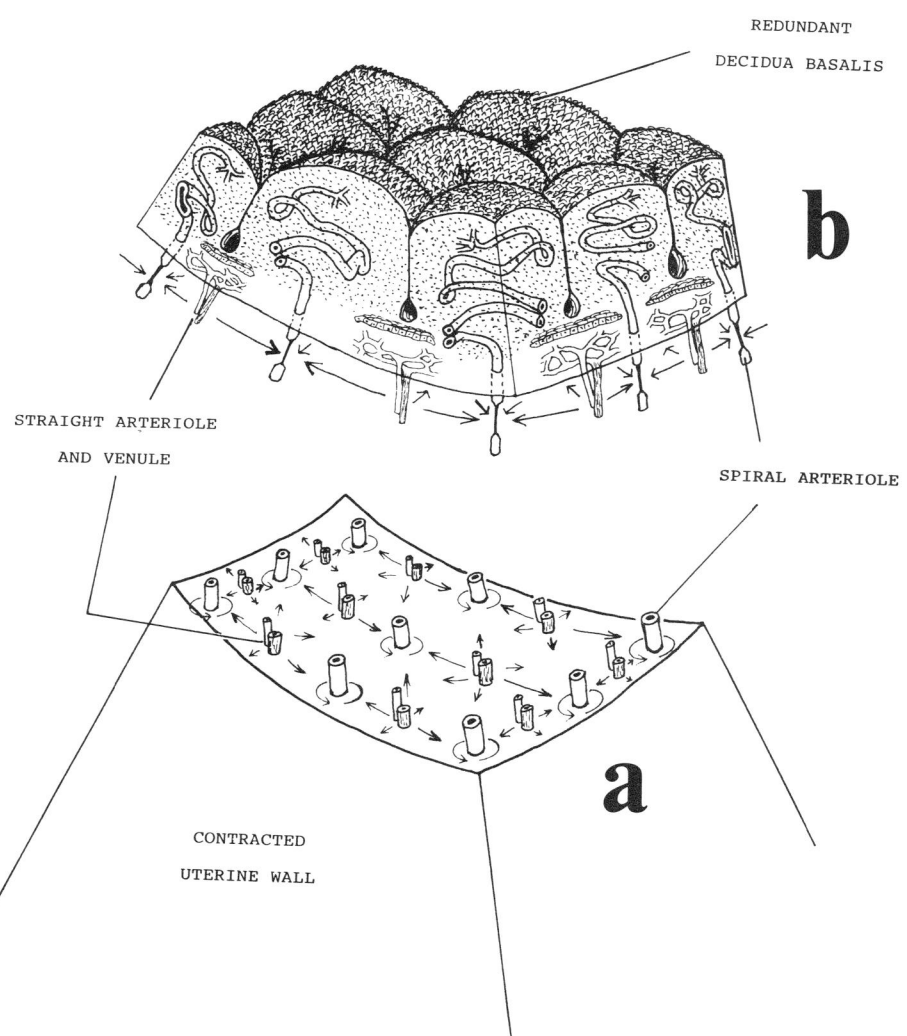

Fig. 8.1. Diagram showing contracted uterine wall (a) and sloughing of endometrial surface (b). The contracted inner muscle layer of the uterus to which the endometrium attaches contains the blood supply to the endometrium (a). "Undermining" reepithelization of the endometrium basalis supplied by the unsphincterized straight vessels has started, and the redundant surface is sloughed (b).

tetanic one to prevent postpartum hemorrhage, a catastrophic form of hemorrhage.*

It does not remain tetanic. Involution continues as before, a shrinkage of uterine musculature with intermittent, superimposed contractions and periods of relaxation. A blushing and blanching of the whole organ probably occurs along with blushing and blanching of the healing endometrium, although it is difficult to observe an intact unanesthetized postpartum uterus *in vivo*. The straight arterioles and venules supplying the basal layer of residual endometrium and fundic gland epithelium continue their circulation during uterine spasm, or squeeze, that produces the endometrial surface ischemia. This circulation to the basal layer of endometrium attached to uterine muscle continues throughout all uterine function at all times—during the menstrual cycle, during pregnancy and labor, and during the puerperium. It never varies. Even the decidual changes in the endometrium do not reach or alter the circulation and characteristics of this endometrial layer.

As this base of residual, preserved endometrium begins to proliferate its stroma and glandular epithelium, capillaries sprout from the ends of the healing spiral arterioles that intermittently open and blanch. Capillary bleeding occurs when the spiral or coiled arterioles intermittently open during healing and capillary surface proliferation. The two capillary and stromal beds unite, the one from the straight basal arterioles and venules, and the other from the coiled arterioles. Although they unite in the healing process, their junction becomes the cleavage plane along which endometrial shear occurs during separation of the placenta and during menstruation (fig. 4.4 and 4.5).

If the globoid postpartum uterus, a 10-centimeter in diameter muscle mass, shows on trans section a 3-centimeter slit in the center as the collapsed and obliterated cavity, the margins or hilar walls appear approximately 3.5 centimeters thick. The thickened anterior and posterior walls containing the greater amounts of contracted muscle are each approximately 5 centimeters in thickness. A sagittal cut reveals the fundic thickness to be even greater with the depth of the obliterated cavity no more than 3.5 to 4 centimeters in length. A rough calculation of the surface area

---

*Postpartum hemorrhage from removal of the ectopic abdominally implanted placenta, where no uterine musculature sphincterizes the arteriolar bloodstream to the placenta, is among the most cataclysmic and exsanguinating hemorrhages known. The hour following delivery of the placenta has been designated the fourth stage of labor. The uterus remains in its tetanic contracted state at this time to prevent postpartum hemorrhage.

of the obliterated cavity covered by endometrium basalis would be something like 24 square centimeters. With the menstruating uterus measuring something like 8 x 4 x 2.5 centimeters its cavity surface measures less than 10 square centimeters. The reduction of uterine cavity surface to one-third less by postpartum involution is remarkable enough. Preservation of its living endometrial covering during rapid and complete involution of the uterine musculature in ten days is even more remarkable.

If the uterus did not continue involution after delivery of the placenta, the endometrial cavity surface would complete its reepithelization and healing probably within twenty-four hours. But rapid progress in further involution during the day following delivery rapidly diminishes uterine mass and cavity surface. Any healed area immediately becomes reduced again in surface area with resultant sloughing (fig. 8.2) and continued healing.

Histologic sections of the healing endometrial surface after delivery have always been confusing by showing all stages of healing, sloughing, and bleeding occurring at the same time but in different areas. The immediate tetanic squeeze and ischemia of the entire placental site, i.e., the entire endometrial surface, gives way after approximately an hour or less to regional areas of healing in which different spiral arterioles at different times relax their sphincters temporarily to supply each area the blood circulation necessary for surface healing and repair. What is taking place is special reduction in the inner muscular layer of spiral arteriole sphincters with convergence of the unchanging endometrium basalis as the uterine cavity surface reduces toward its prepregnancy size. This new surface, like the old, is supplied only by the spiral arterioles (fig. 8.2[b] and [c]), and is the contracting and diminishing surface between the bases of the retained glands in the endometrium basalis. The reducing and reforming endometrial surface projects inward as soon as it reepithelizes forming lochia. In the human, lochia, especially if infected, is passed more than resorbed. This performance of the uterus is a continuation of its involution and emptying function that began with the onset of labor. It occurs on a smaller scale and is therefore brief in the much smaller surfaces of the menstruating uterus. As the larger healing surface of the vascular postpartum uterine cavity is converted to an ischemic volume of dead tissue by uterine involution, it is emptied from the cavity as lochia. In the nonpregnant uterus it is the menstrual flow. As the postpartum uterus approaches the size of the menstruating uterus the two flows, lochia and menstrual, become identical.

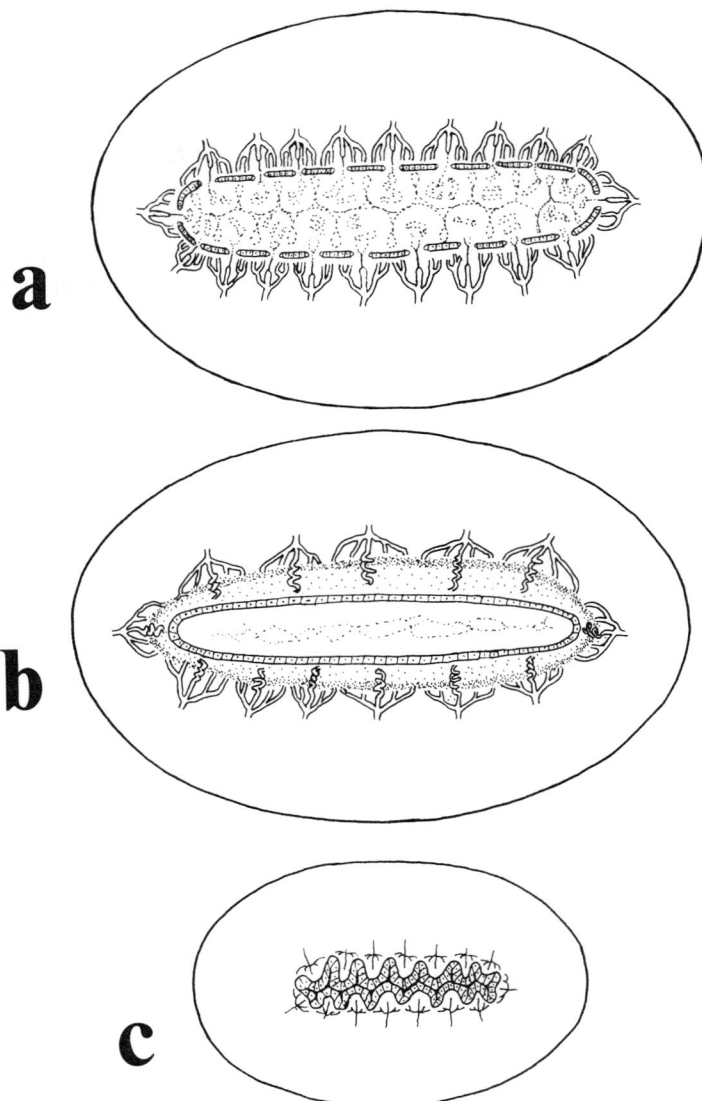

Fig. 8.2. Different stages in the healing and repair of the surface of the endometrial cavity. Reduction of cavity size and surface during postpartum uterine involution (a) causes sloughing of newly formed and redundant endometrial surface. The formation of new endometrial surface (b) over a smaller cavity area and its ischemia and resloughing as the cavity is reduced in area (c) are a result of rapid involution of uterine musculature. The different stages occur simultaneously.

After involution is complete, in approximately ten days, the uterus is relatively tiny with a small cavity surface for handling the mechanical transfer of microscopic egg and sperm particles. Only after involution is complete is the organ prepared for the extensive growth of the next pregnancy. Although involution is the means by which the uterus mechanically opens, hatches, and delivers the egg, clearly the ability of the uterus to reverse the growth of its musculature and size and at the same time preserve its healing and sloughing endometrium during involution is its most remarkable feature. The role of the human uterus in serving the egg its nourishment from its near microscopic size to the comparatively gigantic size it reaches at term is easier to understand than the physiology of the gigantic uterus as it returns to its role of serving the tiny egg at fertilization.

In the human and upper primates, in which the blastocyst attaches immediately to a vascular endometrial layer that shears along its capillary plane, bleeding from the healing vascular membrane occurs in reproductive cycles both with and without egg fertilization.

Part Three
Mechanism of Menstruation

IX
Delivery Performance of the Human Uterus before Term

The normal, or ideal, function of the uterus at term is based upon mechanics that result from the morphology within a fully developed egg at term. "Fully developed egg at term" is a concept clearly accepted but not so clearly defined. When an ideal looking infant with an ideal appearing placenta cries normally at delivery and later nurses ideally, develops, and grows rapidly, it is assumed that the egg reached the term stage of development.

Development of the egg beyond the full term stage of intrauterine development, previously designated as postmature but now designated as post term, is not a concept fully understood. An infant with an excessively large body that still grows within the uterus past the calculated date of confinement obviously will produce some degree of interference with the operation of ideal uterine mechanics during delivery. In these cases, the body mass may be larger in diameter than the fetal head and disturb the normal order of emptying mechanics.

Delivery of the egg before it reaches its full term development, i.e., premature delivery of the egg, is more than a concept. It is a recognized entity that is very real. Although mechanics in premature delivery may also be altered by the different morphology found in earlier stages of egg development, the order of the emptying physiology in the uterus operates the same as it does at term. Premature egg delivery is expressed most often in terms of a time ratio, the weeks of egg development based on full-term development of forty weeks. This ratio is most often determined from assessment of the fetus rather than from assessment of the placenta or from assessment of uterine growth. The entire single, bipolar membrane of the egg is premature in its development, not just its fetal pole, but while prematurity in its development may affect the length of the various stages of uterine involution, it does not affect the pattern and course of emptying physiology. It is the undeveloped morphology of the egg membrane that produces the "apparent" deviation in the course of uterine involution. Only the very first part of uterine involution normally accomplishes opening or hatching of the egg and delivery of the fetus. This can take place in a matter of hours or a day or so at most. In premature delivery, the egg may not be opened but the entire decidua compacta containing the intact egg may be separated and delivered (fig. 9.1). Involution doesn't stop once it starts, irrespective of delivery characteristics, and a small contracted knot constituting the involuted uterus can form behind a small, deceased fetus and placenta obstructed in the vagina. Labor, or involution of the uterus, which is the mechanical emptying force of the uterus, lasts the ten-day period irrespective of the stage to which the pregnancy developed or whether a normal delivery takes place. Involution is not partial, shortened, or arrested; it takes place or it doesn't. There are few, if any, cases on record of a woman carrying a calcified fetus for years with her cervix six centimeters or more dilated from a previously arrested labor. Portions of the uterine musculature may involute while others do not, as in the case of fraternal twins that deliver one month or more apart. The portion of the uterine musculature that does involute, however, does so completely. This indicates a definite local influence of the placenta on the uterine musculature lying directly beneath it.

Involution of the uterus that is slightly premature, i.e., labor that begins two weeks or less before term, for example, may not be easily proved. In such a case delivery mechanics may be normal with no obvious evidence of prematurity in fetus, placenta, or uterus. Since the length of normal pregnancy varies, the time interval from either the last menstrual period or conception date is not very accurate in determining such a minimal premature onset of involution.

Premature involution of the uterus that occurs three weeks or more before term can usually be determined through assessment of the delivered fetus. The delivery mechanics and uterine physiology may appear to be essentially identical to those of a term pregnancy, but the fetus may show evidence of prematurity in either its physical characteristics or its functional adjustment after delivery.

Premature involution four weeks before term may also result in the same mechanics accomplished with an egg developed to term; however, the mother may experience a more difficult and painful course before and during delivery. A multiparous woman who usually experiences little or no pain in labor may report pain during labor only in premature labors, in

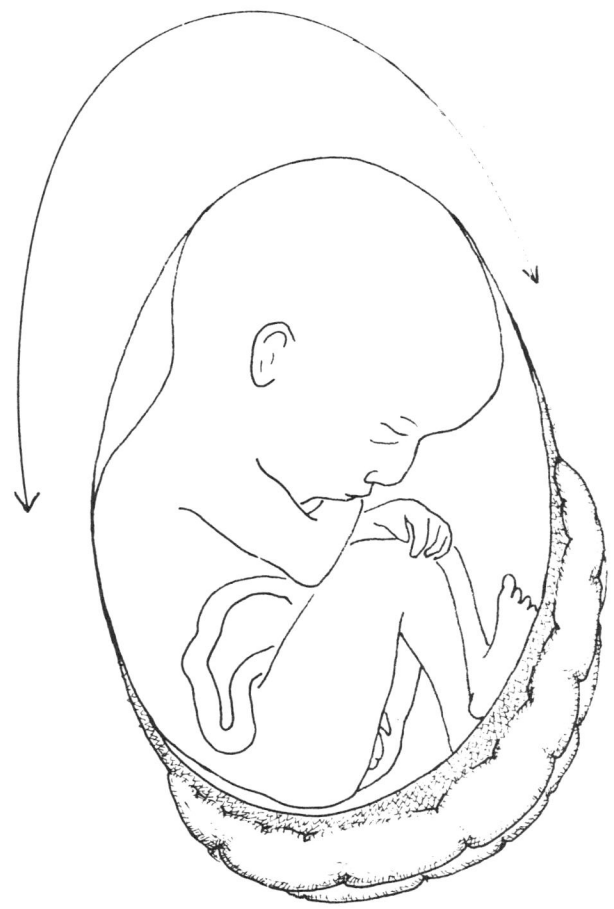

Fig. 9.1. The intact decidua compacta containing the intact premature egg. Although the prematurely developed egg is smaller and may deliver intact as shown here, the egg developed to term rarely if ever does. Uterine forces applied to the larger-term egg opens the resilient egg membranes in their fetal area, shown here bulging from the contracted placenta.

which instance her retrospective report of a more painful labor may be the only indication that uterine involution was slightly premature.

Complications of labor and delivery increase with greater prematurity of involution, roughly in proportion to the degree of prematurity in the egg membrane. The smaller the size of the fetal pole the greater the number of breech, shoulder, and transverse lie presentations that occur. A smaller size of the uterus at the onset of its involution may be discernible, but what may not be discernible in the egg until after delivery is the size of the fetal pole in relation to the surface of placental attachment. Sonography can distinguish the thicker portions of the placental membrane, but even after delivery this ratio is only implied in the contracted placenta. The delivered extrafetal membranes can be filled with water revealing the general locations of the placenta and the opening in the membranes (fig. 9.2). This may be helpful in proving the diagnosis of hemorrhage from the edge of a low-lying placenta during labor. Care must be exercised in utilizing this method, however. Whenever the complete, unopened decidua compacta containing the complete, intact egg membrane is delivered, as represented in figure 9.1 (almost always a case of premature involution with a small fetal pole), the placenta contracts and the extraplacental membranes distend. The volume, surface, and pressure of the fluid remain unchanged after such delivery, but the placenta contracts, correspondingly expanding the bulge of the elastic extraplacental membrane surfaces. Within the uterus the splinting effect from tension of the musculature against the fluid volume that holds the placenta in its stretched and functioning state (see chap. 4) maintains a different ratio of the two surfaces of the egg. The stretched placental surface inside the uterine musculature is a greater proportion, and the extraplacental surface is a lesser proportion of the total egg surface.

In the first half of pregnancy, during formation of the placenta, the ratio of its fixed mechanical surface* to the extraplacental egg surface increases reaching one-third at three months. Early in the last half of pregnancy, after the uterine cavity is obliterated by the enlarging egg, the placenta covers one-half of the egg's surface. As term approaches, the extraplacental membrane surface over the egg expands more rapidly, and the placental area (the villous area) again drops to one-third of the egg surface (Beck, 1942).

---

*The surface against which uterine musculature applies pressure directly, i.e., the villous surface.

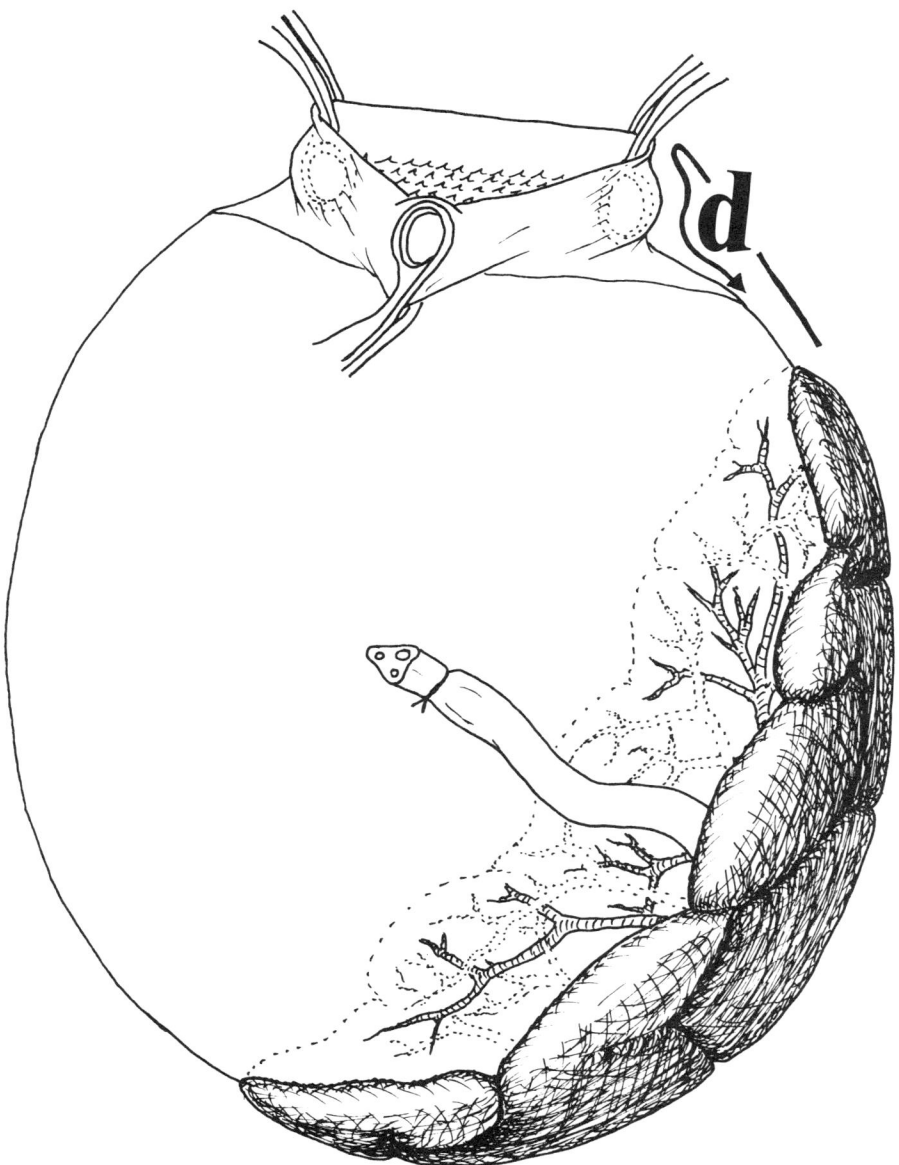

Fig. 9.2. Diagram of delivered extrafetal portion of the membranes filled with water. The relative locations of the placenta and the opening through which the fetal pole was expressed are indicated, with (d) representing the distance of placental edge from cervical opening.

The changing ratio of placental surface from one-half the egg's surface in the middle of pregnancy to one-third at term is accommodated within its decidua compacta enclosure and accounts for the altered application of uterine mechanical force in cases of premature involution of the uterus (fig. 9.3). The uterine mechanics that open the egg membranes become altered by this different morphology of the egg membrane. At term the intact, attached placenta at the beginning of labor blocks involution in the muscularis overlying its third of the uterine cavity surface. Involution may be blocked in the muscularis overlying as much as half the uterine cavity surface when labor occurs before term. With the placenta blocking involution in as much as half the inner muscular layer of the uterine cavity, effacement and dilation of the cervix cannot be completed with the ease of a term fetus and placenta. The circular constricting or binding layer of the musculature cannot shift its pressure to hindwaters behind the fetus without shifting also behind the placenta (compare fig. 9.3 with figs. 5.2, 5.6, and 5.7). The reduced extraplacental cavity surface and smaller fetal surface and fetal volume are too insufficient in ratio to form the ideal wedge or fulcrum for physically and geometrically accomplishing complete cervical dilation. The seal of the decidua capsularis over the internal opening of the cervical canal may not be pulled over sufficiently wide fetal surface to effect rupture of the membranes. The extraplacental egg membrane surface, less than two-thirds of the total egg surface, has insufficient overlying muscularis to thicken enough to reduce cavity volume without the greater likelihood of prematurely separating the placenta. There is also a greater encroachment upon and embarrassment of the uterine circulation to the placenta during an intermittent uterine contraction. The result of premature egg anatomy during labor is separation of the decidua vera frequently without rupture of the membranes. The smaller fetus and smaller but relatively larger placenta form a soft tissue mass, the conceptus, within the endometrial surface which is separated and extruded abruptly and intact through the incompletely dilated cervix (fig. 9.1). The premature egg is delivered, but it is not hatched. Actually it is the endometrium compacta, the fused superficial surfaces of the endometrium, that is separated and delivered, but the solid and enclosed fluid contents of the egg are contained within the separated membrane as in a term pregnancy before labor.

This performance of the uterus during premature involution appears to be a performance quite different from its involution at term, but the difference as far as uterine physiology goes is only apparent, not real in an

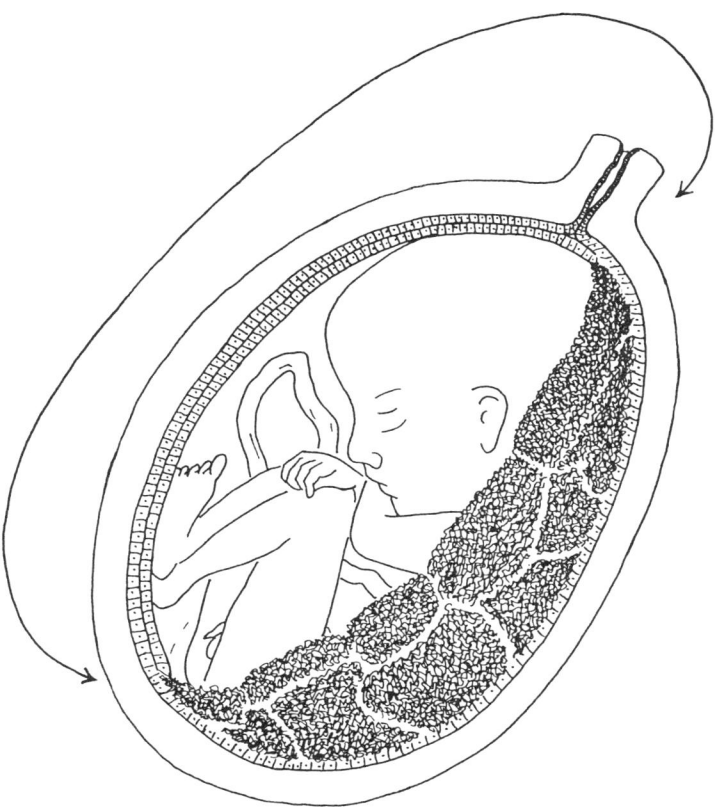

Fig. 9.3. Diagram of a hemisected preterm uterus with the removed portion of the uterus dissected off the intact placenta. The larger placental surface reduces the extraplacental area of the egg surface with overlying musculature and, therefore, hinders the expelling force of the musculature. Compare with figures 5.5 and 5.6.

unstimulated, unmolested, and unmanipulated uterus. If the prematurely involuting uterus is stimulated or manipulated in any way, the organized, orderly sequence of involution that ordinarily empties the uterus can become disorganized. The uterus is apt to overcontract in its circular binding layer, resulting in incarceration of the premature conceptus. What is regarded as the normal emptying mechanics of the uterus is the effect of full-term egg anatomy on involution of uterine musculature. Stages of involution exaggerated or emphasized by term-egg anatomy (rupture of the membranes) are subdued, minimized, and briefly passed over, while still other stages are magnified by early egg morphology, e.g., fetal expulsion. The pattern and order of the uterine emptying functions thus remain unchanged in all stages of pregnancy, and any apparent differences are in mechanical effects from changed egg morphology only. The pattern of all other functions, including endocrine functions, are the same in both premature and term involution.

A conceptus of three to three and a half months with its forming placenta covering only a third of the cavity surface may be delivered, in which the live embryo moves and attempts respiration. Its life functions have been sustained throughout the stages of uterine involution that constituted labor of an abortion by the incompressible fluid held with the embryo within intact egg membranes. In all cases of uterine emptying in pregnancies that do not reach term, involution of the uterus courses through the same steps to completion over the same period of time, approximately ten days.

With this constant ten-day period of involution, the extent of involution that takes place is dependent upon the extent of uterine growth that has previously taken place. Since involution is a reverse of growth in the musculature, it becomes greater the more advanced the gestation. The greater the amount of uterine growth the greater is the size of the egg that produces it and the greater is the extent of uterine emptying involution. The egg, as it grows, by simultaneously stimulating uterine growth and, therefore, enlarging uterine involution, adequately provides and maintains its own delivery function. In normally developed gestations the delivery function developed and maintained within uterine musculature by the egg is no more and no less sufficient than that necessary to separate and deliver an egg the size to which it has grown.

Recognizing and understanding the enormous intimacy of the developing egg and its delivery mechanics is not a trivial matter. It has an evolutionary history that extends back to the origin of mammals and the mammalian uterus. All existing mammals reproduce through the functions of a uterus (see chaps. 1, 2, and 3). In placental mammals this organ provides the mechanical support and movement of a massively developed egg that can grow to 10 percent or more of the mother's weight. It also returns rapidly to an organ with functions capable of supporting and moving a new egg the size of a dust particle. The key function of the mammalian uterus in the reproductive cycle is its capacity or ability to convert itself to structure and function necessary for mechanically handling the mammalian egg at both extremes of its size. Its conversion to a massive organ of one thousand grams through growth that accompanies egg enlargement over the gestation period, as stated in the previous chapter, is much easier to comprehend than its involution, its ten-day conversion to a relatively tiny organ of fifty grams (Beck, 1942).

In cases of extremely premature involution of the uterus that occur in the first four to six weeks of pregnancy, the involution of uterine musculature is free from any mechanical engagement with egg anatomy. The physical surface and volume of the early conceptus is entirely endometrial membrane. In many, if not most, of these early premature uterine involutions the egg has succumbed. Its death, with resultant cessation of further growth stimulus to uterine musculature, is followed by uterine involution and emptying. The endometrial surface may contain the degenerating egg, but the membrane surface containing any and all placental primordium is gathered into a volume, separated, and delivered, often as an endometrial cast (fig. 9.4). After the endometrial cast is passed, involution continues over its ten-day period, during which time the endometrial surface reepithelizes and heals before the next menstrual cycle begins.

Still earlier premature involutions result in what appears to be delayed, heavy, irregular menstrual bleedings with clots. Bits of necrotic endometrial mucous membrane (decidua) may be grossly discernible in the bloody flow. Cramps and dysmenorrhea may be reported by a woman who otherwise experiences no discomfort during her menses and in whom a previous pregnancy test may have been positive. The ten-day period of involution reduces the slightly enlarged uterus to its premenses size. In these most extreme premature involutions of the uterus, the pregnancy terminates at some stage shortly after the egg implants within the endometrium. The endocrine influence of the implanted egg advances endometrial growth, development, and enlargement beyond its premenstrual stage. It accomplishes this indirectly through the ovary, as the developing egg itself at this tiny stage of development is too small to produce

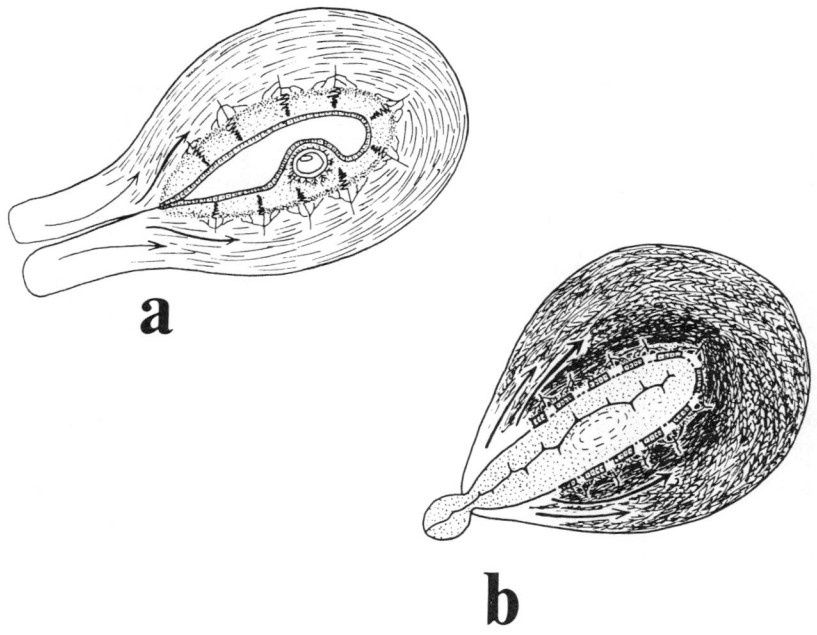

Fig. 9.4. Delivery of an early intrauterine pregnancy of only a few weeks. It consists of delivery of only the endometrial surface decidua. The egg, present (a) or blighted and absent (b), offers no mechanical influence at this early stage. The mass expelled is only the redundant endometrium undermined by regenerating epithelium beginning to repair and cover the intact basalis.

sufficient blood levels of estrogen and progesterone necessary to maintain and advance growth of the much larger uterus. Any retained fragments of endometrium or clots in these early premature involutions (micromiscarriages or microabortions) may offer some slight physical and mechanical obstruction to the emptying involution, just as retained placental fragments may do to a much larger extent in the emptying function from later stages of pregnancy. Irregular bleeding may result until the uterine cavity is completely emptied and its surface reepithelized and healed. Involution is complete and the cavity becomes empty when the surface layer of endometrium, the spreading and thickening tissue surface designed to contain the enlarging egg during pregnancy, is separated, passed, and its area of attachment, its base, is healed over.

As long as the uterine cavity surface is reducing in area, uterine involution is in progress with healing and reepithelization repeated for each increment of surface reduction. Repetitious cavity surface reduction and its healing after delivery at term are extensive because of the large surface and size of the involuting uterine cavity. Cavity surface reduction and healing are much less extensive in cases of microabortions, because the uterus and its cavity have enlarged very little from the size they attain during the menstrual cycle. Microabortions, which are cases of involution and emptying of the uterus very shortly after the egg implants, are the earliest most premature forms of labor recognizable in accordance with conventional concepts and terminology. Irregular bouts of "dysmenorrheal" cramps, associated with heavier than usual amounts of "menstrual" bleeding and passage of mucous debris with clots, are known to occur unassociated with pregnancy. Pathologists have long avoided recognizing the decidual reaction in passed tissue fragments as a basis for diagnosing pregnancy, preferring the conclusive evidence of trophoblastic villi instead. Some uterine physiology such as uterine growth may be identical in both irregular menstrual periods and microabortion, but uterine involution and emptying may not be adequate or present at all in cases of irregular menstrual cycles. Sampling of the endometrium in the latter case may or may not show progestational effects indicative of ovulation.

When ovulation has not occurred, the endometrium remains in the estrogenic, or growth phase. It remains the slightly expanded membrane in its growth stage, still attached at its base. Intermittent uterine contractions without involution damage the surface of the endometrium but do not separate it. The blood supply to the surface is not permanently interrupted. When ovulation occurs, the growing musculature, primed for involution by progesterone (from the corpus luteum in the ovary), shrinks or involutes after the unfertilized egg fails to implant. The endometrial surface is separated by the labor and delivery of this involution, in which contraction of the involuting uterus clamps off the blood supply to the membranous placental site even though no placenta has started to form. The period of endometrial ischemia is severe and of such length during uterine involution to kill the surface endometrium. The shrinking, involution, and contraction of all uterine musculature, which includes sphincter musculature about the bases of the spiral arterioles but not the straight vessels of the endometrial base, empties the uterus by compression and obliteration of any and all volume or space that previously constituted the uterine cavity. Healing of the reduced endometrial surface undermines its adhering necrotic surface layer and causes it to bleed and separate as the living tissue below is reepithelized. This is the smallest scale of the uterine performance, which becomes greatly magnified in a term delivery.

In the case of irregular menstrual bleeding without ovulation, the myometrium without involution, like the endometrium, remains slightly enlarged in its growth phase. The next menstrual cycle applies its growth stimulation to a uterus in which the slightly enlarged musculature and the slightly expanded endometrial surface of the previous cycle have not been reduced by involution and resultant cavity emptying. What follows is intermittent progressive enlargement of the uterus and uterine cavity surface in steps (if ovulation fails to occur along the way). This persistent uterine growth in spurts is brought about from the sequence of newly initiated but never completed (anovulatory) menstrual cycles. The enlarged and expanded endometrial surface is retained and, without the physical and mechanical stretch and distention from placenta formation in a growing egg, becomes redundant. Its normally enlarging and expanding concave surface becomes convex and grows inward toward the center of the cavity, into an ever diminishing area, toward a spherical center of zero surface and volume. During a cramp or intermittent uterine contraction, the redundant membrane is gathered into a compacted volume illustrated in figure 9.5[a]. The redundant endometrium is compressed and compacted and then released without its separation or passage. The unobstructed but damaged endometrial blood vessels continue to supply the damaged endometrial surface that bleeds more profusely the more it is damaged and the more it grows. Menstrual cycles are repeatedly initiated. Their uterine effects—repeated growth stimuli—become superimposed, stacked one on top of another. The uterus gradually enlarges and the endometrial vessels

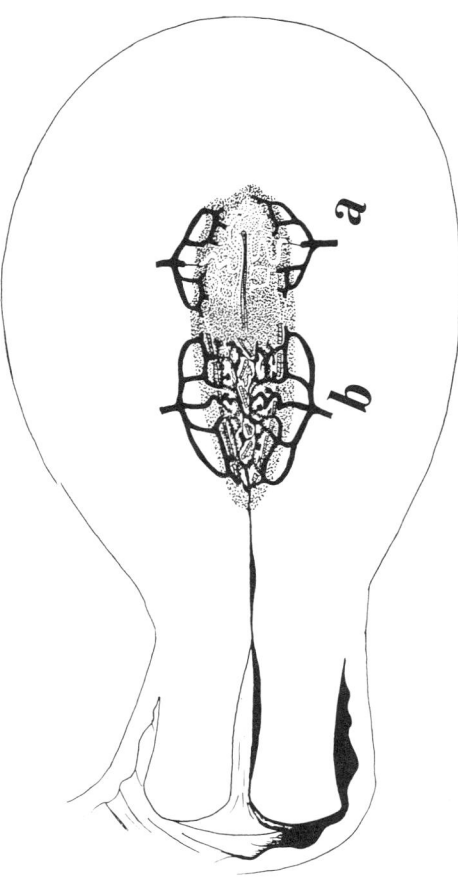

Fig. 9.5. Two stages of the delivery functions of the uterus presented in figure 9.4 but in the absence of an implanted egg (menstruation). The endometrial ischemic phase (a) is followed by endometrial slough and healing (b).

continue to bleed more profusely. The endometrial hemorrhage in each cycle represents the opened maternal circulation of an absent placenta that was destined and designed for the blighted egg responsible for that particular surge of endometrial growth.

Once ovulation occurs progestational effects result in involution and cavity emptying with passage of the proliferated endometrial surface. The uterus and endometrium not only shrink in size, but the spiral vessels supplying the separating endometrial surface are clamped off until the new endometrial membrane, a smaller surface, begins to form. Bleeding is reduced and controlled. Each menstrual cycle is thus physiologically ended. The well-established and familiar histologic picture of progestational endometrium mirrors the far more important and critical progestational effect within the myometrium, the ability of the musculature to involute.

In the normally menstruating woman the period of uterine growth and enlargement, the estrogenic phase, is balanced by a corresponding period of involution and shrinkage that keeps the uterus an average uniform size. The growth phase is variable in the length of its time span, but the involutional phase is a constant ten-day time interval. By lengthening the growth phase, the involuting phase is rendered latent, and pregnancy thus physiologically prolongs the uterine growth and enlargement phase of the human reproductive cycle.

Full comprehension of human reproductive physiology exceeds the limits of the concepts imposed by conventional terminology. Implantation of a fertilized egg within the endometrium is a stage in human pregnancy and pregnancy functions that is understood and is often diagnosed easily. Preimplantation stages of pregnancy, the stage of the morula and the stage of the free blastocyst, are stages of pregnancy that are not as easy to diagnose *in vivo* and when diagnosed are usually rare diagnoses accidentally established in histologic material or in experimental conditions of harvesting human eggs for *in vitro* fertilization. No endocrinologic test is presently available that can conclusively detect the human, unattached fertilized egg *in vivo*. There is some question in accordance with present concepts and understanding whether such an endocrinologic test can ever be developed. Entrance of a sperm into egg cytoplasm does not produce change in maternal physiology. It is the detection of serum and blood levels of metabolic products from the tiny endometrium-fixed or attached egg that is the basis of early detection in the human. In lower placental mammals the egg grows to a larger size before it attaches to the endometrium and thus becomes capable of producing detectable levels of substances. Before it becomes fixed in the uterus the egg in lower placental mammals is apparently sustained by ovarian control of the uterus that prevents uterine emptying and abortion. The human egg in its unimplanted stage, as well as its early implanted stage, is too small to produce the necessary estrogen and progesterone levels required by the uterus in those early stages of pregnancy. When the tiny egg implants or embeds in the endometrium it continues the blood levels of estrogen and progesterone supplied from the ovary by rapidly producing a rising titer of ovary-stimulating hormone. Detection of this hormone substance is the basis of most pregnancy tests. The early embedding egg, while too small to produce mechanical effects in the uterus, prolongs the uterine growth phase of its particular reproductive cycle by physiologically becoming a replacement of the pituitary's role in the reproductive function. As the human egg enlarges into mostly its placenta at first, it begins to produce all the endocrines necessary to advance the pregnancy functions within the uterus that were previously initiated from the ovary. The developing egg in human pregnancy eventually replaces the endocrine physiology of all other endocrine glands that control uterine function.

Premature labor and emptying of the uterus before the fertilized egg embeds in the endometrial surface cannot occur according to conventional concepts, as the earliest premature labor physiologically possible in humans would be in a pregnancy that ends in a normal menstrual period. This is not the conventional concept of menstruation. Instances of every recognizable stage of pregnancy are known to end with labor and delivery functions of the uterus. If pregnancy is considered to begin with conception or fertilization of the egg in mammals, then establishment of pregnancy functions in the uterus prior to pregnancy as defined must be recognized. A human egg, fertilized or unfertilized that does not embed, is a product of physiology that ends in labor and delivery of its uterine bed, conventionally considered to be a normal menstrual period. Labor and delivery of the earliest embedded egg is recognized easily enough as an early abortion, and an unfertilized egg that never lives to embed can be recognized as the cause of normal menstrual periods. The fertilized egg that dies just before or while embedding is not recognized as the labor of abortion that it becomes by conventional concept and definition.

The progestational phase of the menstrual cycle is the phase that prepares the uterine musculature for involution and emptying before and irrespective of whether egg fertilization and embedding ever take place.

The intimacy of egg production and its delivery physiology is obvious. If the ovary produces an egg, it also produces that egg's delivery mechanism, as the physiology that produces the egg is also the physiology that establishes its delivery mechanics in the uterus. If an egg is not produced and ovulated from the ovary, no delivery mechanism is completed in the uterus.

Because a single egg cell in the ovary is too small at the beginning of the reproductive cycle to produce the amount of hormone necessary for the preparation of its delivery mechanism in an organ the size of the uterus, growth and development of many egg cells in the ovary are necessary to initiate the cycle. All except one are sacrificed in the human (atresia) when one egg cell becomes large enough to produce the hormone level necessary to complete its delivery physiology in the uterus. The single produced egg (ovulation) only occurs after its complete delivery functions are fully established in the uterus and not before. Based on the average twenty-eight-day menstrual cycle with the first day of the cycle beginning with the first day of the menstrual period, fertilization of the egg is considered more likely, on an average, to occur as early as the tenth or as late as the seventeenth day. In three to four days of the menstrual cycle the egg traverses tube and uterine cavity. In another three to four days or so it implants and grows to the stage of producing a level of APL sufficient to preserve and extend corpus luteum function. Effects of the unfertilized egg may be considered to begin about the twenty-third to the twenty-fifth day. The failing support of the corpus luteum from the pituitary gland begins corpus luteum regression with dropping estrogen and progesterone levels. Involution in the uterus beginning on or about the twenty-fourth day and lasting its ten days carries the uterus through its healing and bleeding phases of the menstrual period, which actually marks the beginning of the ensuing cycle rather than the end of the preceding cycle. A growth stimulus enters and overlaps the end of an involution period in the uterus. Involution is usually complete when the growth surge occurs, relaxing the uterus and the sphincterized, coiled arterioles that open upon the damaged endometrial surface and hemorrhage. Involution and growth attain an equal balance in the uterus of a normally menstruating woman. When ovulation does not take place, increase of one over the other results in either increase or decrease in uterine size. The final tissue to involute in uterine involution is the inner sphincter layer of the musculature pierced by the spiral arterioles.

The delivery, or emptying, function of pregnancy is the primary function of the mammalian uterus and is completely established in the uterus with ovulation early in each reproductive cycle. This uterine function produces the uterine responses of the estrous cycle and is the dominant physiology that has evolved in the mammalian reproductive cycle to include the hatching functions of the large mammalian egg in placental mammals. It advances to the menstrual cycle that has evolved in upper primates, in which the ovarian egg has functionally become still smaller and entirely hematrophic in its nutrition.

In the mammalian lineage, egg opening, or hatching functions, became a phylogenetic addition to uterine function at the beginning of involution. As the egg developed completely intramembranously and totally hemotrophically in the evolution of primates, blood-vessel clamping, or sphincterization, of the arterial blood supply to the membrane surface—the maternal placenta—became a necessary addition at the end of delivery mechanics. The climax of involution and contraction is the uterine cramp that produces the endometrial ischemia. In infertile cycles this dominating cramping accounts for primary dysmenorrhea and usually disappears as the uterus relaxes and the new growth cycle gets under way.

The bleeding of menstruation is the result of merging nutritive functions, initiated at the beginning of the reproductive cycle in the uterus of upper primates, with those dominant mechanical functions of the upper primate uterus that mechanically empty the uterus and end the cycle. Initial reproductive functions are increase in vascularity and endometrial growth that begin by ending the involution period of the preceding cycle. Menstruation results from overlap of this double simultaneous response. When the time between menstrual periods lengthens and menstrual periods become spaced farther and farther apart, bleeding becomes less and finally ceases. The overlap of adjacent cycles spreads apart, separating each cycle and giving the endometrium time to completely heal before it begins the growth of the next cycle. Conversely, when the time between menstrual periods shortens, as in anovulatory cycles, the overlap of adjacent cycles is increased, the involution period shortens or ceases, and growth and bleeding become heavy and continuous.

Bibliography

Allen, E. 1923. Racial and familiar cyclic inheritance and other evidence from the mouse concerning the cause of oestrus phenomena. *Am. J. Anat.* 32:293–304.
Allen, W. M. 1931. 1. Cyclical alterations of the endometrium of the rat during the normal cycle, pseudo pregnancy and pregnancy. 2. Production of deciduomata during pregnancy. *Anat. Rec.* 48:65–103.
Amoroso, E. C. 1959. Comparative Anatomy of the placenta. *Annals New York Academy of Sciences.* 75:855–70.
Arey, L. B. 1942. *Developmental Anatomy.* 4th ed. Philadelphia: Saunders.
Barcroft, J., and D. H. Barron. 1942. Circulation in the placenta of the sheep. *J. Physiol.* 100:20.
Barcroft, J., W. Herkel, and S. Hill. 1933. Rate of blood flow and gaseous metabolism of the uterus during pregnancy. *J. Physiol.* 77:194–206.
Bartelmez, G. W. 1933. Histological studies on the menstruating mucous membrane of the human uterus. *Carnegie Inst. Pub. 24; Contrib. to Embryol.* 142:141–85.
———. The mechanism of menstruation. *Anat. Rec.* 97:380.
———. 1937. Menstruation. *Physiol. Review.* 17:28–72.
———. 1941. Menstruation. *JAMA.* 116:702.
Bourliere, F. 1956. *Mammals of the World.* London: George G. Harrap Co.
Caldwell, H. 1884. On the arrangement of the embryonic membranes in marsupial animals. *Qtr. J. M. Sci.* 25:655–58.
Corner, George W. 1923. Oestrus, ovulation and menstruation. *Physiol. Rev.* 3:457–81.
———. 1927. Relation between menstruation and ovulation in the monkey; Its possible significance for man. *JAMA.* 89:1838–40.
———. and A. E. Amsbaugh. 1917. Oestrus and ovulation in swine. *Anat. Rec.* 12:287–91.
Crossen, H. S., and R. J. Crossen. 1944. *Diseases of Women.* 9th ed. St. Louis: C. V. Mosby Company.
Csapo, A. I. 1954. The molecular basis of myometrial function and its disorders. *La Prophylaxic en gynealogic et obstetrique.* Geneva: 694.
———, and B. A. Resch. 1979. Induction of preterm labor in the rat by the antiprogesterone. *Am. J. Ob. Gyn.* 134:823–27.

Cunningham, F. G., P. C. Mac Donald, and N. F. Grant. 1989. *Williams Obstetrics.* 18th ed. Norwalk, Conn./San Mateo, Calif.: Appleton & Lange.

Curtis, A. H. 1947. *Textbook of Gynecology.* 5th ed. Philadelphia: Saunders.

Daron, G. H. 1936. The arterial pattern of the tunica mucosa of the uterus in *Macacus rhesus. Am. J. Anat.* pp. 349–420.

Dobzhansky, Theodosius 1962. *Mankind Evolving.* New Haven, Conn.: Yale University Press.

Doggett, T. H. 1961. The biologic foundation of menstruation and uterine bleeding. *The Journal of the Florida Medical Association.* 47:1011–13 (March).

Dukes, H. H. 1955. *The Physiology of Domestic Animals.* Ithaca, N.Y.: Comstock Publishing Associates, a division of Cornell Univ. Press.

Elder, J., and R. M. Yerkes. 1936. The sexual cycle of the chimpanzee. *Anat. Rec.* 67:119–43.

Engle, E. T. *Menstruation and Its Disorders.* Springfield, Ill.: Charles C. Thomas.

Flexner, L. B., and Gellhorn. 1942. The comparative physiology of placental transfer. *Am. J. Ob. and Gyn.* 43:965.

Fluhman, C. F. 1936. The estrin-deprivation theory of menstruation. *Endocrin.* 20:318.

Gillespie, E. C., E. M. Ramsey, and S. R. M. Reynolds. 1949. The pattern of uterine growth during pregnancy in monkeys. *Am. J. Ob. and Gyn.* 59:949–59.

Goodman, L., and G. B. Wislocki. 1935. Cyclical uterine bleeding in a new world monkey *(Ateles geoffroyi). Anat. Rec.* 61:379.

Grant, V. 1963. *The Origin of Adaptations.* New York: Columbia Univ. Press.

Grosser, O. 1909. *Vergleichende Anatomie und Entwicklungsgeschichte der Eihäute und der Placenta.* Wein: Wilhelm Braumüller.

Hamlett, G. W. D. 1939. Reproduction in American monkeys. 1. Estrous cycle, ovulation and menstruation in cebus. *Anat. Rec.* 73:171.

Hartman, C. 1916. Studies in the development of the oposum, *Didelphys virginiana.* 5. The phenomena of parturition. *Anat. Rec.* 19:251–60.

———. 1923 (a). The oestrous cycle in the opossum. *Am. J. Anat.* 32:353–421.

———. 1923 (b). Breeding habits, development and birth of the opossum. In the Appendix to the 1921 Report of the Sec. of the Smithsonian Institute. Washington.

———. 1929. The homology of menstruation. *JAMA.* 92:1992–95.

———. 1930. The corpus luteum and the menstrual cycle together with the correlation between menstruation and implantation. *Am. J. Ob. and Gyn.* 19:511.

———. 1931. *JAMA.* 97:1863.

———. 1932. Studies in the reproduction of the monkey, *Macacus rhesus,* with special reference to menstruation and pregnancy. *Contrib. to Embryol. No. 433, Pub. 134.* Carnegie Inst. 23:1–161.

Hartree, A. S. 1989. Multiple forms of pituitary and placental gonadotrophins. *Oxford Review of Reproductive Biology* 11:147–177.

Heape, W. 1901. The sexual season of mammals and the relation of the pro-estrus to menstruation. *Qtr. J. Micro. Sci.* 44:1.

Herbst, Arthur L., Mischell, Stenchever, and Drodgemueller. 1992. *Comprehensive Gynecology.* 2nd ed. St. Louis: Mosby-Year Book, Inc.

Heuser, C. H. 1927. A study of the implantation of the ovum of the pig from the stage of the bilaminar blastocyst to the completion of the fetal membranes. *Carnegie Inst. Contrib. to Embryol.* 106:229–42.

Hill, J. P. 1897. The placentation of *Perameles.* Contributions to the embryology of the Marsupialia 1. *Qtr. J. Micro. Sci.* 40:385–446.

———. 1900. Contributions to the embryology of Marsupialia 2 and 3. *Qtr. J. Micro. Sci.* 43:1–22.

———. 1911. The early development of the Marsupialia with special reference to the native cat *Dasyurus viverrinus.* Contributions to the embryology of the Marsupialia 4. *Qtr. J. Micro. Sci.* 56:1–134.

Hubrecht, A. A. W. 1889. Studies in mammalian embryology 1. The placentation of *Erinaceus europaeus,* with remarks on the phylogeny of the placenta. *Qtr. J. Micro. Sci.* 30:283.

———. 1908. Early ontogenetic phenomena in mammals. Chapter 5 on placentation. *Qtr. J. Micro. Sci.* 53:98–171.

Hulse, F. S. 1961. Warfare, demography, and genetics. *Eugenics Quart.* 8:185–97.

———. 1964. The paragon of animals. *Eugenics Quart.* 11:1–10.

———. 1968. Migration and cultural selection in human genetics. *The Anthropologist.* Special volume: 1–21. Univ. of Delhi (India).

Jolly, A. 1972. *The Evolution of Primate Behavior.* New York: The Macmillan Company.

Long, J. A., and H. M. Evans. 1922. The oestrous cycle in the rat and its associated phenomena. *Memoirs of Univ. of Calif.* 6. Berkeley: Univ. of California Press.

Markee, J. E. 1940. Menstruation in intraocular endometrial transplants in the rhesus monkey. *Carnegie Inst. Pub. 518 Contrib. to Embryol.* 28:220–308. Washington.

Mossman, J. W. 1937. Comparative morphogenesis of the fetal membranes and accessory uterine structures. *Contrib. to Embryol.* 158:129–246.

Nobak, C. R. 1939. The changes in the vaginal smears and associated cyclic phenomena in the lowland gorilla, *Gorilla Gorilla. Anat. Rec.* 73:209–20.

Novak, E. 1921. *Menstruation and Its Disorders. Gynecological and Obstetrical Monographs.* New York: D. Appleton and Co.

Poulton, E. B. 1884. The structures connected with the ovarian ovum of Marsupialia and Monotremata. *Qtr. J. Micro. Sci.* 24:118–28.

Papanicolaou, G. N. 1923. Oestrus in mammals from a comparative point of view. *Am. J. Anat.* 32:285–92.

Pendle, G. 1954. *Paraguay.* London: Institute of International Affairs.

Ramsey, E. M. 1947. The vascular pattern of the endometrium of the pregnant rhesus monkey (abstract). *Anat. Rec.* 93:363.

Reynolds, S. R. M. 1947. The physiologic basis of menstruation; a summary of current concepts. *JAMA.* 135:552.

———. 1949. *Physiology of the Uterus.* 2nd ed. New York: Paul B. Hober.

Roberts, S. J. 1971. *Veterinary Obstetrics and Genital Diseases (Theriogenology).* Ithaca, N. Y.: Published by the author. Distributed by Edwards Brothers, Inc., Ann Arbor, Michigan.

Romanes, G. J. 1901. *Darwin and after Darwin.* 3rd ed. Chicago: Open Court.

Sanders, P. M. 1988. A fossil reptile embryo from the middle Triassic of the Alps. *Science.* 239:780–82.

Science. March 1988. *Research News.* 239:1091–92.

Scott, J. R., P. J. DiSaia, C. B. Hammond, and W. N. Spellacy. 1990. *Danforth's Obstetrics and Gynecology.* 6th ed. Philadelphia: Lippincott.

Sharman, G. B. 1970. Reproductive physiology of marsupials. *Science.* 167:1221.

Simpson, G. G. 1944. *Tempo and Mode in Evolution.* New York: Columbia Univ. Press.

———. 1953. *Life of the Past—an Introduction to Paleontology.* New Haven: Yale Univ. Press.

———. 1967. *The Meaning of Evolution.* Rev. ed. New Haven: Yale University Press.

Smith, O. W., and G. S. Smith. 1946. Studies concerning the cause and purpose of menstruation. *J. Clin. Endocr.* 6:483.

Smuts, B. B. et al. 1987. *Primate Societies.* Univ. of Chicago Press.

Stockard, C. R. 1923. Introductory; the general morphological and physiological importance of the oestrous problem. *Am. J. Anat.* 32:227–83.

Storer, T. J., R. L. Usinger, R. C. Stebbins, and J. W. Nybakken. 1972. *General Zoology.* 5th ed. New York: McGraw-Hill.

Turner, W. 1897. Some general observations of the placenta with especial reference to the theory of evolution. *J. Anat. and Physiol.* 11:33–53.

Wislocki, G. B. 1929. On the placentation of primates with a consideration of the phylogeny of the placenta. *Carnegie Inst. Contrib. to Embryol.* 20:51–80.

———. 1930. On a series of placental stages of a platyrrhine monkey, *Ateles Geoffryi*, with some remarks upon age, sex, and breeding period in platyrrhines. *Carnegie Inst. Contrib. to Embryol.* 22:173–92.

Young, L. B. (ed.) et al. 1970. *Evolution of Man.* New York: Oxford Univ. Press.